Mechanism Design for Robotics

Mechanism Design for Robotics

Special Issue Editors

Marco Ceccarelli
Alessandro Gasparetto

MDPI • Basel • Beijing • Wuhan • Barcelona • Belgrade

MDPI

Special Issue Editors
Marco Ceccarelli
University of Rome Tor Vergata
Italy

Alessandro Gasparetto
Università degli Studi di Udine
Italy

Editorial Office
MDPI
St. Alban-Anlage 66
4052 Basel, Switzerland

This is a reprint of articles from the Special Issue published online in the open access journal *Robotics* (ISSN 2218-6581) from 2018 to 2019 (available at: https://www.mdpi.com/journal/robotics/special_issues/MDR)

For citation purposes, cite each article independently as indicated on the article page online and as indicated below:

LastName, A.A.; LastName, B.B.; LastName, C.C. Article Title. *Journal Name* **Year**, *Article Number*, Page Range.

ISBN 978-3-03921-058-9 (Pbk)
ISBN 978-3-03921-059-6 (PDF)

Contents

About the Special Issue Editors

Marco Ceccarelli received his Ph.D. in Mechanical Engineering from La Sapienza University of Rome, Italy, in 1988. He is a Professor of Mechanics of Machines at the University of Rome Tor Vergata, Italy, where he chairs LARM2: Laboratory of Robot Mechatronics. His research interests cover the subjects of robot design, mechanism kinematics, experimental mechanics with special attention to parallel kinematics machines, service robotic devices, mechanism design, and history of machines and mechanisms whose expertise is documented by several published papers in the fields of Robotics. He has been a visiting professor in several universities around the world and since 2014 at the Beijing Institute of Technology. He is an ASME fellow. Professor Ceccarelli serves on several Journal Editorial Boards and conference Scientific Committees. He is an Editor of the Springer book series on Mechanism and Machine Science (MMS) and History of MMS. Professor Ceccarelli is the President of IFToMM, the International Federation for the Promotion of MMS. He has started several IFToMM-sponsored conferences, including MEDER (Mechanism Design for Robotics) and MUSME (Multibody Systems and Mechatronics).

Alessandro Gasparetto obtained a PhD in Applied Mechanics at the University of Brescia (1996). Since 2007 he has been a Professor of Mechanics Applied to Machines at the University of Udine. The main topics of his scientific research are dynamic modeling, experimental analysis and control of flexible link mechanisms, and robotics and mechatronics for advanced applications. To date, his scientific research has produced around 200 scientific publications. He is a scientific reviewer for more than 60 international journals, and for numerous national and international bodies. He has participated in various national and international research projects. He is the Chair of the Permanent Commission for History of Mechanism and Machine Science (HMMS) of the IFToMM (International Federation for Promotion of Mechanism and Machine Science). Since 2013 he has been the President of IFToMM Italy, the Italian section of IFToMM.

robotics

MDPI

Editorial

Mechanism Design for Robotics

Marco Ceccarelli [1],* and Alessandro Gasparetto [2]

[1] Laboratory of Robot Mechatronics, Department of Industrial Engineering; University of Rome Tor Vergata, Via del Politecnico 1, 00133 Roma, Italy

[2] Polytechnic Department of Engineering and Architecture, Università degli Studi di Udine, Via delle Scienze 206, 33100 Udine, Italy; alessandro.gasparetto@uniud.it

* Correspondence: marco.ceccarelli@uniroma2.it

Received: 17 April 2019; Accepted: 17 April 2019; Published: 19 April 2019

MEDER 2018, the IFToMM International Symposium on Mechanism Design for Robotics, was the fourth event of a series that was started in 2010 as a specific conference activity on mechanisms for robots. The first event was held at Universidad Panamericana de Ciudad de Mexico, Mexico, in September 2010; the second was held in 2012 at Beihang University in Beijing, China; the third one was held in 2012at Aalborg University in Denmark; and the fourth one was organized at Udine University in Italy.

The aim of the MEDER Symposium is to bring researchers, industry professionals, and students together from a broad ranges of disciplines dealing with mechanisms for robots, in an intimate, collegial, and stimulating environment. Again, in the 2018 MEDER event, we have received significant attention regarding this initiative, as can be seen by the fact that the Proceedings contain contributions by authors from all around the world.

The Proceedings of MEDER 2018 Symposium have been published within the Springer book series on MMS, and the book contains 52 papers that have been selected after review for oral presentation. These papers cover several aspects of the wide field of robotics dealing with mechanism aspects in theory, design, numerical evaluations, and applications.

This Special Issue in *Robotics* (https://www.mdpi.com/journal/robotics/special_issues/MDR) has been obtained as a result of a second review process and selection, but all the papers that have been accepted for MEDER 2018 are of very good quality with interesting contents that are suitable for journal publication, and the selection process has been difficult.

The papers in this Special Issue are focused on the design of novel mechanisms for robots, and on the development of new methodologies for the modeling and control of robotic devices. In particular, a couple of papers deal with walking robots, namely, one presents a novel humanoid robot with parallel architectures [1] and the other one is focused on a hexapod robot [2]. Several papers are focused on teleoperation and HRI (human–robot interactions) to deal with the design of a safe teleoperated system for Doppler sonography [3] and of a safety mechanism for cobot joints [4], the development of techniques for hand guidance of a mobile robot [5], and the use of artificial muscles in haptic interfaces [6]. Other papers address problems regarding the design of specific actuators for robots, such as underactuated fingers [7], compliant rotary actuators [8], and robotic wrists [9]. There are also three papers on special robotic systems for a robot for artistic painting [10], a cylindrical rolling robot [11], and a cable-driven robot [12]. The last paper of this Special Issue presents a methodology for the design of optimal robotic trajectories [13].

The members of the International Scientific Committee for MEDER Symposium are gratefully acknowledged for helping us to achieve the success of MEDER 2018 and this Special Issue:

Marco Ceccarelli (Univ. of Cassino, Italy, Chair)

Ding Xilun (Beihang Univ., China)

Grigore Gogu (SIGMA Clermont, France)

Mario Acevedo (Univ. Panamericana, Mexico)

Shaoping Bai (Aalborg Univ., Denmark)

Yukio Takeda (Tokyo Inst Tech., Japan)

Alba Perez (Idaho State Univ., USA)

Yan Jin (Belfast Univ., UK)

Erwin Lovasz (Politehnica Univ. Timisoara, Romania)

Said Zeghloul (Poitiers Univ., France)

Victor Petuya (Univ. of Basque Country, Spain).

We thank the authors, who have contributed very interesting papers on several subjects, covering many fields of "Mechanism Design for Robotics", and for their cooperation in revising papers on time in agreement with the reviewers' comments. We are grateful to the reviewers for the time and effort they spent in evaluating the papers to a very tight schedule, which permitted the publication of this Special Issue.

We thank the Udine University in Udine, Italy, for having hosted the MEDER 2018 event with a very efficient local organising committee. We would like to thank the sponsorship of IFToMM (International Federation for the Promotion of Mechanism and Machine Science), IFToMM Italy, the IFToMM Technical Committees of Linkages and Mechanical Controls, and *Robotics* and *Mechatronics*, which have supported the conference event, in which more than 15 countries participated.

We would like to thank the publisher and editorial staff of the *Robotics* for accepting and helping with the publication of this Special Issue, a process that began in 2016.

We are grateful to our families, since without their patience and comprehension it would not have been possible for us to organise MEDER 2018, IFToMM International Symposium on Mechanism Design for Robotics and this Special Issue.

References

1. Russo, M.; Cafolla, D.; Ceccarelli, M. Design and Experiments of a Novel Humanoid Robot with Parallel Architectures. *Robotics* **2018**, *7*, 79. [CrossRef]

2. LI, R.; Meng, H.; Bai, S.; Yao, Y.; Zhang, J. Stability and Gait Planning of 3-UPU Hexapod Walking Robot. *Robotics* **2018**, *7*, 48. [CrossRef]

3. Arévalo, J.S.S.; Laribi, M.A.; Zeghloul, S.; Arsicault, M. On the Design of a Safe Human-Friendly Teleoperated System for Doppler Sonography. *Robotics* **2019**, *8*, 29. [CrossRef]

4. Ayoubi, Y.; Laribi, M.A.; Zeghloul, S.; Arsicault, M. V2SOM: A Novel Safety Mechanism Dedicated to a Cobot's Rotary Joints. *Robotics* **2019**, *8*, 18. [CrossRef]

5. Weyrer, M.; Brandstötter, M.; Husty, M. Singularity Avoidance Control of a Non-Holonomic Mobile Manipulator for Intuitive Hand Guidance. *Robotics* **2019**, *8*, 14. [CrossRef]

6. Franco, W.; Maffiodo, D.; de Benedictis, C.; Ferraresi, C. Use of McKibben Muscle in a Haptic Interface. *Robotics* **2019**, *8*, 13. [CrossRef]

7. Zappatore, G.A.; Reina, G.; Messina, A. A Toolbox for the Analysis of the Grasp Stability of Underactuated Fingers. *Robotics* **2019**, *8*, 26. [CrossRef]

8. Kittinanthapanya, R.; Sugahara, Y.; Matsuura, D.; Takeda, Y. Development of a Novel SMA-Driven Compliant Rotary Actuator Based on a Double Helical Structure. *Robotics* **2019**, *8*, 12. [CrossRef]

9. Shah, D.; Wu, Y.; Scalzo, A.; Metta, G.; Parmiggiani, A. A Comparison of Robot Wrist Implementations for the iCub Humanoid. *Robotics* **2019**, *8*, 11. [CrossRef]

10. Scalera, L.; Seriani, S.; Gasparetto, A.; Gallina, P. Non-Photorealistic Rendering Techniques for Artistic Robotic Painting. *Robotics* **2019**, *8*, 10. [CrossRef]

11. Puopolo, M.; Jacob, J.; Gabino, E. Locomotion of a Cylindrical Rolling Robot with a Shape Changing Outer Surface. *Robotics* **2018**, *7*, 52. [CrossRef]

12. Boschetti, G.; Carbone, G.; Passarini, C. Cable Failure Operation Strategy for a Rehabilitation Cable-Driven Robot. *Robotics* **2019**, *8*, 17. [CrossRef]
13. Boscariol, P.; Richiedei, D. Trajectory Design for Energy Savings in Redundant Robotic Cells. *Robotics* **2019**, *8*, 15. [CrossRef]

robotics

MDPI

Article

Stability and Gait Planning of 3-UPU Hexapod Walking Robot

Ruiqin Li [1,*], Hongwei Meng [1], Shaoping Bai [2], Yinyin Yao [1] and Jianwei Zhang [1,3]

[1] School of Mechanical Engineering, North University of China, Taiyuan 030051, China;
menghw@st.nuc.edu.cn (H.M.); Yao_yinyin@st.nuc.edu.cn (Y.Y.); zhang@informatik.uni-hamburg.de (J.Z.)
[2] Department of Mechanical and Manufacturing Engineering, Aalborg University, 9220 Aalborg, Denmark;
shb@mp.aau.dk
[3] Department of Informatics, University of Hamburg, 22527 Hamburg, Germany
* Correspondence: liruiqin@nuc.edu.cn; Tel.: +86-351-3921300

Received: 26 July 2018; Accepted: 29 August 2018; Published: 31 August 2018

Abstract: The paper presents an innovative hexapod walking robot built with 3-UPU parallel mechanism. In the robot, the parallel mechanism is used as both an actuator to generate walking and also a connecting body to connect two groups of three legs, thus enabling the robot to walk with simple gait by very few motors. In this paper, forward and inverse kinematics solutions are obtained. The workspace of the parallel mechanism is analyzed using limit boundary search method. The walking stability of the robot is analyzed, which yields the robot's maximum step length. The gait planning of the hexapod walking robot is studied for walking on both flat and uneven terrains. The new robot, combining the advantages of parallel robot and walking robot, has a large carrying capacity, strong passing ability, flexible turning ability, and simple gait control for its deployment for uneven terrains.

Keywords: hexapod walking robot; 3-UPU parallel mechanism; kinematics; stability; gait planning

---

1. Introduction

Legged robots have advantages over the wheeled robots in their flexibility of movement and the adaptability of the environment. Legged robots can adapt to different terrains. A multi-DOF design enables it to actively adjust the height of the body according to the operational requirements to ensure the balance and stability of the body.

To date, many legged robots have been developed [1–4]. The NOROS robot developed by Ding et al. [2,3] is a modular walking robot. It consists of up to eight autonomous leg modules, each equipped with various sensors. A self-recovery approach is applied by imitating the self-recovery motion of insects. Ma et al. [4] proposed a robot which consists of six legs of the same structure distributed evenly on a platform. Each leg of the robot has a structure similar to that of a telescopic parallelogram mechanism with folding capability, which is beneficial to the gait planning and real-time control of the robot. The parallel type mobile robot formed by parallel mechanism applied to the legged robot can greatly improve the load/weight ratio of the mobile robot, reduce the energy consumption, and prolong the walking time. Our interest is to develop a hexapod walking robot that utilizes a parallel mechanism, namely, a three-DOF translational 3-UPU parallel mechanism.

The concept of 3-UPU parallel mechanism was first introduced by Tsai [5], which can produce three-dimensional translation movement. Gregorio [6] presented a 3-UPU parallel manipulator, named 3-UPU wrist for spherical motions with its prismatic pairs actuated. Ji et al. [7] proposed a 3-UPU translational parallel robotic manipulator with an equal offset in its six universal joints and obtained 16 solutions for forward kinematics. Han et al. [8] analyzed kinematic sensitivity of the 3-UPU parallel mechanism. Huang et al. [9] studied feasible instantaneous motions and kinematic characteristics of an orthogonal 3-UPU parallel mechanism. Wu et al. [10] designed a configuration-switch mechanism for 3-UPU parallel mechanism.

The kinematics and dynamics analyses of 3-UPU parallel mechanisms have been well developed in literatures [11–15]. Two novel 3-UPU parallel kinematics machines with two rotations and one translation are proposed in literature [13]. In addition, singularity [14,15] and stiffness [16] of the 3-UPU parallel mechanism are also studied.

There are some reports about applications of parallel mechanism in mobile robots. A rolling biped robot was proposed based on a 3-UPU parallel mechanism [17]. Gait and stability analyses were presented and four rolling modes of the mechanism were discussed and simulated. Gu [18] proposed to apply a typical 3-UPU parallel mechanism in a quadrupedal walking robot. Wang et al. [19] presented a bipedal locomotor consisting of two identical 3-DOF tripod leg mechanisms with a parallel manipulator architecture. Sugahara et al. [20] presented a design of a battery driven bipedal robot, which used 6-DOF parallel mechanism for its each leg. Wang et al. [21] proposed a quadruped/biped reconfigurable parallel legged walking robot, in which the robot can change between biped and quadruped walking modes according to real-time road conditions. Each leg is composed of a 3-UPU parallel mechanism. The robot walks as a quadruped generally and changes to bipedal walking when walking up and down stairs.

Walking robots have to walk following a certain pattern of leg movement, namely, the gaits. Hirakoso et al. [22] developed a multi-legged gait prototype robot with four legs consisted of redundant joint. An optimal control system was proposed to control any motion for the four-legged robot with redundant joint. Sun et al. [23] proposed a transformable wheel-legged mobile robot that integrated stability and maneuverability of wheeled robot and obstacle climbing capability of legged robot. When under wheeled mode, the robot avoids the obstacle using a motion control strategy that combines three basic cases of translation, rotation, and arc motion, while the robot climbs the obstacle by legs. A method of free gait generation that utilizes the primary/secondary gait for both straight line and circular body trajectories was proposed by Bai [24]. Four constraints of primary gait were discussed. When the walking machine cannot move using the primary gait, the secondary gait is generated to adjust the leg position and enable the vehicle to keep on moving. Gait planning combined with path planning was also developed [25,26].

Gong et al. [27] focused on the dynamic gaits control for complex robot with 20 DOFs. Using composite cycloid to plan the swing foot trajectory curve made the velocity and acceleration to be zero which can reduce the impact of collision with ground and energy loss. Li et al. [28] presented a three-dimensional model of a quadruped robot which has six DOFs on torso and five DOFs on each leg. Matsuzawa et al. [29] proposed a crawling motion to reduce the risk of malfunction due to falling when a legged robot travels across rough terrain.

Winkler et al. [30,31] developed trajectory planning for legged locomotion that automatically determines the gait sequence, step timings, footholds, swing-leg motions, and six-dimensional body motion over rough terrain, without any additional modules. Neunert et al. [32] proposed a trajectory optimization framework for whole-body motion planning through contacts. Zhao et al. [33] developed a motion generation approach for a hexapod robot Octopus-III to control the robot to walk along the planned trajectory. The approach coordinates the body motion and the feet motions to fulfill requirements of walking stability and kinematic feasibility simultaneously. Oliveira et al. [34] studied locomotion patterns of hexapod robots, including metachronal wave gait, and tetrapod and tripod gaits.

Up to date, the reported research works of legged robots are mainly focused on biped, quadruped, and hexapod robots. Most of these robots have independent driving joints on the legs, which requires many actuators. In the case of a single leg of three joints, a hexapod robot needs at least 18 motors which greatly increases the weight of the robot. Moreover, the position of each foot has to be considered, which adds complexity to control.

The paper presents a hexapod walking robot designed with a 3-UPU parallel mechanism, aiming to reduce the number of driving motors and improve load capacity and terrain adaptability. The new design is characterized by alternating motion between the two platforms of the 3-UPU parallel mechanism to realize robot walking. The parallel mechanism is used as both an actuator to generate walking and also a connecting body to connect two groups of three legs. The new walking robot is thus able to walk with simple gait by very few motors.

The paper is organized as follows. Firstly, the configuration of the hexapod walking robot is presented in Section 2. The forward and inverse kinematic solutions of the 3-UPU parallel mechanism are derived in Section 3. The workspace of the robot is obtained using joint constraint conditions and inverse kinematics equation in Section 4. The movement stability of the robot is analyzed using the center of gravity projection method in Section 5. The gait of the robot is planned under the conditions of ensuring the efficiency of the walking mode of the robot in Section 6. A case study is included in Section 7, and the work is concluded in Section 8.

2. Configuration of 3-UPU Hexapod Walking Robot

A 3-UPU hexapod walking robot is proposed, as shown in Figure 1a. The 3-UPP mechanism, shown separately in Figure 1b, connects two platforms. On each platform, three retractable legs, distributed evenly on the triangular platforms, are mounted to support the robot. Servo motors are mounted at the end of each supporting leg to adjust its length in fitting uneven ground. A separate pressure sensor is installed on each supporting leg to detect whether the supporting leg is in contact with the ground.

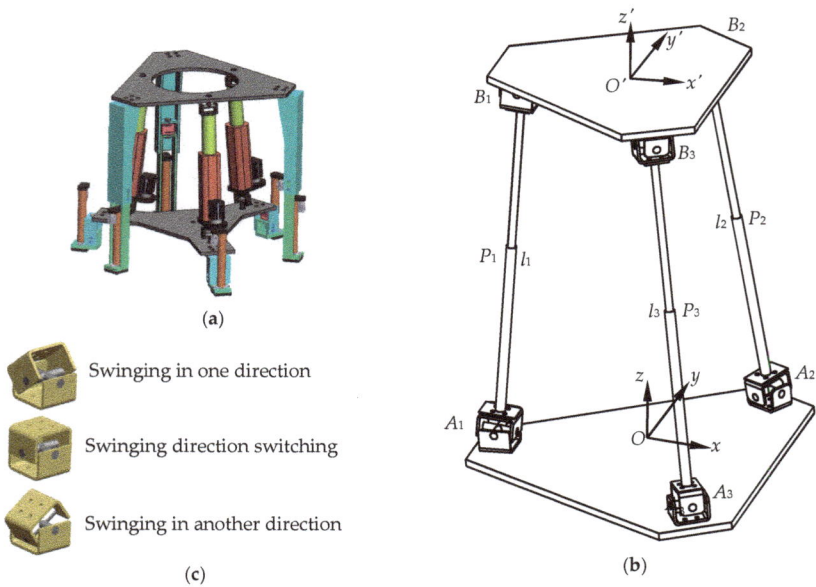

Figure 1. Structure of 3-UPU hexapod walking robot. (**a**) 3D model. (**b**) 3-UPU parallel mechanism. (**c**) The orientation of the Hooke joint.

The upper and lower platforms of the 3-UPU parallel mechanism are connected by three limbs. The structures of the Hooke joints in three limbs are different. In two of these limbs, the structures of the Hooke joints are the same, which can supply two rotational DOFs. In another limb, two forks of the Hooke joint are just in contact with each other, as shown in Figure 1c. The active fork can only swing in one direction during the movement of the robot. The swinging direction can be switched when the active fork swings to its highest position. This structure of the Hooke joint can effectively control the moving direction of 3-UPU hexapod walking robot.

The rotating shafts on forks fixed to the platform are tangent to the circumcircle of the equilateral triangle formed by mounting positions of three Hooke joints. Because there are three couples which limit the three rotational DOFs of the mechanism, thus the moving platform has only three translational DOFs.

3. Kinematics of 3-UPU Parallel Mechanism

We briefly describe the kinematics of 3-UPU parallel mechanism, upon which the locomotion kinematics is developed.

3.1. Inverse Kinematics Solution

The moving, or the upper platform of 3-UPU parallel mechanism is a mechanism with only three-dimensional translations. The inverse kinematics problem of 3-UPU parallel mechanism is to solve the lengths (l_1, l_2, l_3) of the equivalent driving links for given position (x, y, z) of the moving platform.

As shown in Figure 1b, the moving coordinate system O'-$x'y'z'$ is connected to the moving platform. The origin O' is located at the center of the moving platform. x' axis is coincident with $O'B_1$, and point B_1 is located at the negative direction of x' axis. y' axis is parallel to line B_3B_2. z' axis is perpendicular to the upper platform and pointing away from the mechanism. The coordinate system O-xyz is connected to the lower, or the static platform. The origin O is located at the center of the static platform. The x axis is coincident with OA_1, and point A_1 is located at the axis' negative part. y axis is parallel to line A_3A_2. z axis is perpendicular to the static platform and the direction is upward. Moreover, the circumradii of the moving platform and the static platform are noted as r and R, respectively. The equivalent link lengths of the limb i are noted by l_i ($i = 1, 2, 3$).

According to the geometric characteristics of the mechanism, the closed vector loop equation is established as:

$$A_iB_i = -OA_i + OO' + O'B_i, (i = 1, 2, 3) \tag{1}$$

$$|A_iB_i| = l_i, (i = 1, 2, 3) \tag{2}$$

Let the coordinate of the geometric center O' in the static coordinate system O-xyz be (x, y, z). The vector OO' can be expressed as:

$$OO' = \begin{bmatrix} x, & y, & z \end{bmatrix}^{\mathrm{T}} \tag{3}$$

and OA_i and $O'B_i$ ($i = 1, 2, 3$) can be expressed as:

$$
\begin{aligned}
OA_1 &= \begin{bmatrix} -R, & 0, & 0 \end{bmatrix}^{\mathrm{T}} & O'B_1 &= \begin{bmatrix} -r, & 0, & 0 \end{bmatrix}^{\mathrm{T}} \\
OA_2 &= \begin{bmatrix} \tfrac{1}{2}R, & \tfrac{\sqrt{3}}{2}R, & 0 \end{bmatrix}^{\mathrm{T}} & O'B_2 &= \begin{bmatrix} \tfrac{1}{2}r, & \tfrac{\sqrt{3}}{2}r, & 0 \end{bmatrix}^{\mathrm{T}} \\
OA_3 &= \begin{bmatrix} \tfrac{1}{2}R, & -\tfrac{\sqrt{3}}{2}R, & 0 \end{bmatrix}^{\mathrm{T}} & O'B_3 &= \begin{bmatrix} \tfrac{1}{2}r, & -\tfrac{\sqrt{3}}{2}r, & 0 \end{bmatrix}^{\mathrm{T}}
\end{aligned} \tag{4}
$$

From Equations (1), (3), and (4), the following expressions can be obtained.

$$
\begin{aligned}
A_1B_1 &= \begin{bmatrix} x + \Delta, & y, & z \end{bmatrix}^{\mathrm{T}} \\
A_2B_2 &= \begin{bmatrix} x - \tfrac{1}{2}\Delta, & y - \tfrac{\sqrt{3}}{2}\Delta, & z \end{bmatrix}^{\mathrm{T}} \\
A_3B_3 &= \begin{bmatrix} x - \tfrac{1}{2}\Delta, & y + \tfrac{\sqrt{3}}{2}\Delta, & z \end{bmatrix}^{\mathrm{T}}
\end{aligned} \tag{5}
$$

where $\Delta = R - r$.

According to Equation (2), the position inverse solution can be obtained.

$$\begin{cases} (x + \Delta)^2 + y^2 + z^2 = l_1{}^2 \\ \left(x - \frac{1}{2}\Delta\right)^2 + \left(y - \frac{\sqrt{3}}{2}\Delta\right)^2 + z^2 = l_2{}^2 \\ \left(x - \frac{1}{2}\Delta\right)^2 + \left(y + \frac{\sqrt{3}}{2}\Delta\right)^2 + z^2 = l_3{}^2 \end{cases} \tag{6}$$

When the basic geometric dimensions of the 3-UPU parallel mechanism and the position (x, y, z) of the moving platform are known, the variation of the equivalent link length l_i (i = 1, 2, 3) of three limbs, i.e., the displacement of three prismatic pairs can be obtained using Equation (6). Similarly, when the upper platform is a static platform and the lower platform is a moving platform, the static coordinate system O-xyz is connected to the upper platform, while the moving coordinate system O'-$x'y'z'$ is connected to the lower platform. The coordinate directions do not change, and the closed vector loop equation is established as:

$$\boldsymbol{B_i A_i} = -\boldsymbol{OB_i} + \boldsymbol{OO'} + \boldsymbol{O'A_i}, \ (i = 1, 2, 3) \tag{7}$$

Similarly, the inverse position solution of the upper platform can be obtained as:

$$\begin{cases} (x - \Delta)^2 + y^2 + z^2 = l_1{}^2 \\ \left(x + \frac{1}{2}\Delta\right)^2 + \left(y + \frac{\sqrt{3}}{2}\Delta\right)^2 + z^2 = l_2{}^2 \\ \left(x + \frac{1}{2}\Delta\right)^2 + \left(y - \frac{\sqrt{3}}{2}\Delta\right)^2 + z^2 = l_3{}^2 \end{cases} \tag{8}$$

3.2. Kinematics Forward Solution

Giving the lengths (l_1, l_2, l_3) of the prismatic pairs of three limbs, we need to solve the spatial position (x, y, z) of the moving platform. The solutions of position (x, y, z) can be obtained numerically using Equation (6).

4. Workspace of the 3-UPU Parallel Mechanism

4.1. Factors Influencing the Workspace of the 3-UPU Parallel Mechanism

The reachable workspace of the 3-UPU parallel mechanism is under the influence of several factors: the shortest and longest distance of the limb and the range of rotation angle of Hooke joint.

4.1.1. Shortest and Longest Distances of the Limbs

The position (x, y, z) of the reference point of the moving platform is constrained by the equivalent link length l_i (i = 1, 2, 3) of the three limbs. The three prismatic pairs in the limbs act as driving inputs, which makes the moving platform move in the workspace range. When the limb is the shortest or the longest, servo motor I, which acts as the prismatic pair, is in the zero position or the limit position. The reference point of the moving platform reaches the boundary of the workspace.

Taking the upper platform as the moving platform, the equivalent link length l_i (i = 1, 2, 3) of each limb satisfy the constraint:

$$l_{\min} < l_i < l_{\max} \tag{9}$$

where $l_{\min}$ and $l_{\max}$ are the minimum and maximum link lengths of the limb.

4.1.2. Rotation Angle Range of the Hooke Joint

The Hooke joints are used to connect the limbs to the upper and lower platforms. As shown in Figure 2, the rotation angle θ_1 of the Hooke joint connected to the limb is related to the length l_1 of the limb and the y coordinate of the moving platform. The rotation angle θ_1 should satisfy:

$$\theta_1 = \arcsin \frac{B_1 E}{l_1} = \arcsin \frac{|y|}{l_1} \tag{10}$$

The rotation angle θ_2 of the Hooke joint connected to the static platform is related to the projection length of the limb l_1 at O-xz plane and the z coordinate of the moving platform. The rotation angle θ_2 should satisfy:

$$\theta_2 = \arccos \frac{EF}{A_1 E} = \arccos \frac{z}{\sqrt{l_1{}^2 - y^2}} \tag{11}$$

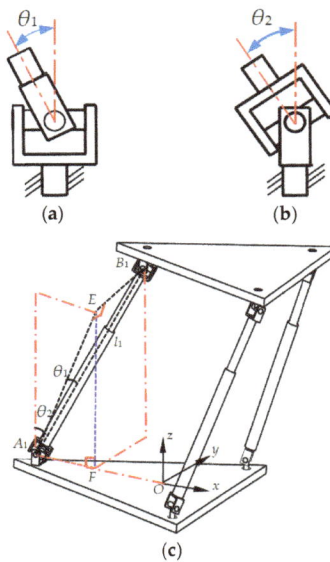

Figure 2. Rotation angles of the Hooke joint in the 3-UPU parallel mechanism. (a) Angle θ_1 of the Hooke joint. (b) Angle θ_2 of the Hooke joint. (c) Rotation angles of the Hooke joint.

4.2. Method for Determining Workspace of the 3-UPU Parallel Mechanism

The workspace of the 3-UPU parallel mechanism is determined using the limit boundary search method. The idea is that when the position of the moving platform is known, the equivalent link length l_i (i = 1, 2, 3) of each limb can be calculated by the inverse solution equation of the mechanism position. The rotation angles of Hooke joint can be solved using Equations (10) and (11). Then, these results are compared with the corresponding limit values, respectively. If any of these values exceed the allowable range, that is, the reference point is outside the workspace. The solutions are

(1)   Defining the structure parameter of the mechanism, including: the circumradii r and R of the upper and lower platforms, the maximum travel range of the equivalent link length l_i (i = 1, 2, 3) of limb i, the maximum rotation angle of Hooke joint.

(2)   Defining the range of the coordinates (x, y, z) of the reference points of the moving platform, or, the search space.

(3) The reference point coordinate is substituted into the position inverse solution equation of the mechanism, the equivalent link length l_i (i = 1, 2, 3) and the Hooke joint angles θ_1 and θ_2 in the limbs are obtained. The results are checked whether they are within the allowable range. If they are, these points are recorded.

(4) The set of points satisfying the condition is the workspace of the 3-UPU parallel mechanism when a given range of values has been searched.

The design parameters are given in Table 1. According to workspace search flow chart in Figure 3, the workspaces of the 3-UPU parallel mechanism are obtained using MATLAB.

Table 1. Design parameters of the 3-UPU parallel mechanism.

Parameter	Symbol and Unit
Equivalent circumradius of the upper platform	r/mm
Equivalent circumradius of the lower platform	R/mm
Range of the equivalent link length of limb i	$l_i (i = 1,2,3)$/mm
Range of the rotation angle of the Hooke joint	$\theta_i (i = 1,2)$/rad

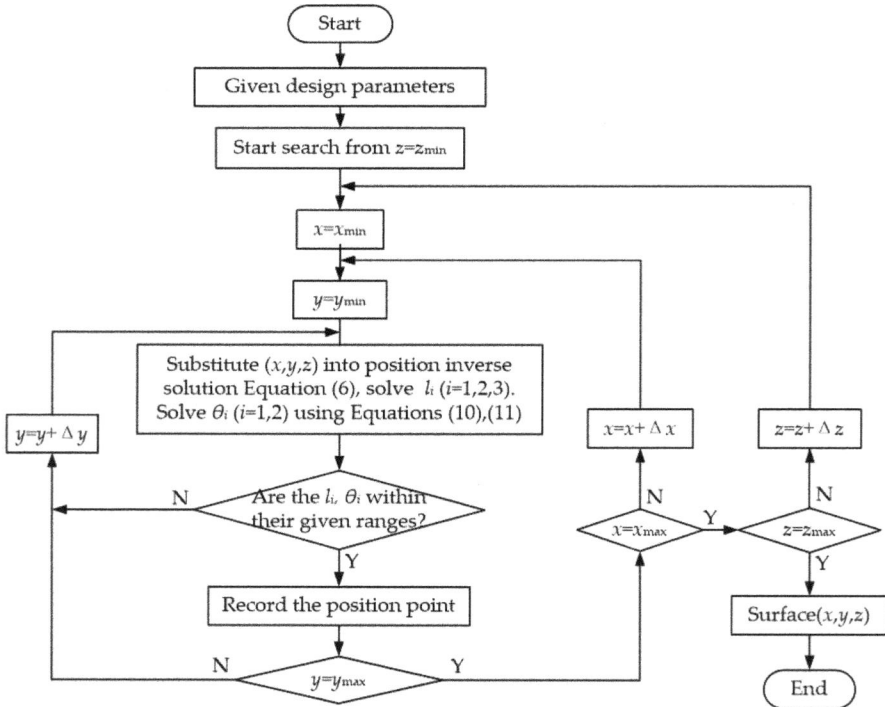

Figure 3. Workspace search flow chart for the 3-UPU parallel mechanism.

5. Stability of 3-UPU Hexapod Walking Robot

The walking of 3-UPU hexapod robot is realized by alternative shifting between the upper and lower platforms, each with three retractable legs to support the whole robot. During the movement of the robot, the robot is supported by three legs and six legs alternately. An overturn can happen in the phase of three-leg support. Stability analysis is thus needed.

Stability of walking robots can be analyzed in terms of static and dynamic stability. A walking robot is statically stable if the horizontal projection of its center of gravity lies inside the support polygon, while it is dynamically stable when the robot walking is stable even if the static stable condition is not satisfied. As the 3-UPU hexapod robot walks in a relative slow speed, we analyze only its static stability.

As shown in Figure 4, three points—E, F, H—are the landing positions of the three supporting legs of lower platform, respectively. They form a supporting triangle. Let the center of gravity of the upper platform and its three legs be marked as G_1, the center of gravity of the lower platform and its three legs as G_2. The center of gravity of the whole robot, located on the line connecting two points G_1 and G_2, is marked as G_3. Figure 5 shows the relation between the projection of the center of gravity and stability margin d. When the projection of the robot's center of gravity G_3 on the ground is located within the supporting triangle, the walking is stable. Otherwise, the walking of the robot is unstable and the robot is likely overturned.

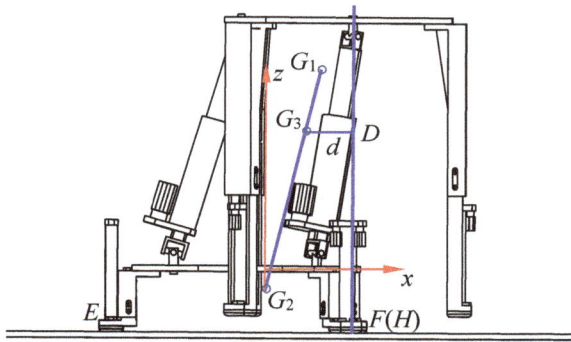

Figure 4. Three centers of gravity considered in the stability analysis.

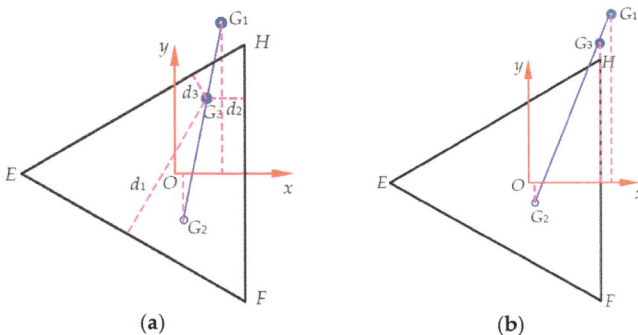

Figure 5. Relation between the projection of the center of gravity and stability margin. (**a**) The projection of the center of gravity is located within the supporting triangle. (**b**) The projection of the center of gravity is located on the boundary of the supporting triangle.

The projections of points G_1 and G_2 are coincident in the initial status, that is the six legs of the robot land on the ground. When the projection of the center of gravity G_3 moves inside the supporting triangle, the amount of movement allowed is the smallest in the direction perpendicular to the boundary. Thus, the displacement of the projection of the center of gravity G_3 in the O-xy plane is analyzed. As shown in Figure 5a, let the horizontal coordinate of points G_1 and G_3 be x_{G1} and x_{G3}, respectively, the mass of the upper platform and its three legs be m_1, the mass of the lower platform and its three legs is m_2, the moment equilibrium equation is established as

$$x_{G1}m_1 - x_{G3}(m_1 + m_2) = 0 \tag{12}$$

To sum up, the horizontal coordinate of the point G_3 is

$$x_{G3} = \frac{x_{G1}m_1}{m_1 + m_2} \tag{13}$$

When the displacement of the upper platform is equal to zero, the projections of the two points G_1 and G_2 are coincident, too. If G_1 is located outside the supporting triangle, as shown in Figure 5b, the robot will lose its stability. The maximum step length of the robot is marked as x_G. Then, the step length of the robot, marked as S, can be expressed as:

$$S \leq x_G \tag{14}$$

Here, the maximum step length of the hexapod walking robot is calculated and analyzed in theory only according to the instability condition. The movement of the platform is restricted by the equivalent link length l_i ($i = 1, 2, 3$) of the limb, which may not reach the maximum step length before the instability.

6. Gait Planning of 3-UPU Hexapod Walking Robot

For the new walking robot, we develop a special type of gait, which is simple and easy to implement.

6.1. A New Gait

The walking of 3-UPU hexapod robot is realized by alternating motion between two platforms. A gait called 3–3 gait, i.e., alternate tripod support gait, is used in the process of walking. The legs are divided into two groups, in turn in the support phase or suspending phase state.

As the robot uses two groups of legs to walk, the duty factor β is defined as 0.5 for the hexapod walking robot. This ensures that the robot has three legs on the ground to keep the body stable, while the robot has relatively high walking efficiency and moderate velocity. The gait diagram is displayed in Figure 6, where the thick line stands for the state when a leg is supporting on the ground and a thin line for legs in the air.

Figure 6. 3–3 gait diagram of 3-UPU hexapod walking robot.

The gait of one cycle in walking is divided into two stages:

Stage I: The six legs of the robot land on the ground in the initial state. To begin with, a group of legs—namely, legs 1, 3, 5—in the lower platform serves as a support, the upper platform lifts, as shown in Figure 7a; Then, the upper platform moves forward, as shown in Figure 7b; Finally, a group of legs, namely—legs 2, 4, 6—in the upper platform lands on the ground as shown in Figure 7c.

Stage II: When a group of legs in the upper platform lands on the ground, the legs in the lower platform—namely, legs 1, 3 and 5—will be lifted, as shown in Figure 7d; Then the lower platform moves forward, as shown in Figure 7e; Finally, these legs land on the ground, as shown in Figure 7f. A cycle of this gait is completed.

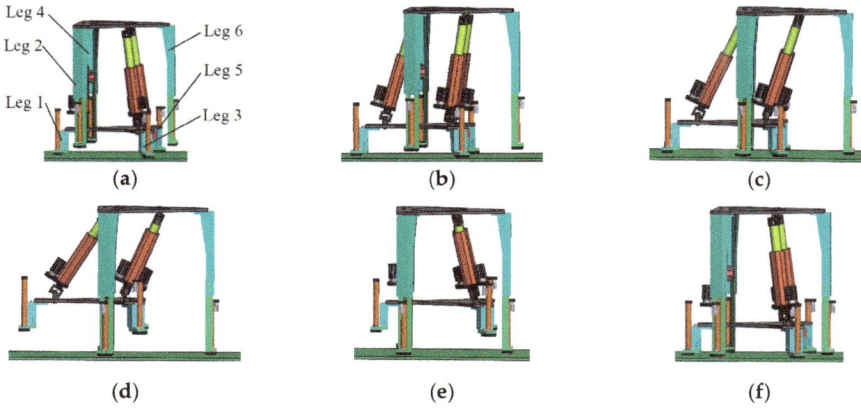

Figure 7. Walking gait of 3-UPU hexapod walking robot. (**a**) Lifting of the upper platform. (**b**) Moving forward of the upper platform. (**c**) Landing on the ground of the upper platform. (**d**) Lifting of the lower platform. (**e**) Moving forward of the lower platform. (**f**) Landing on the ground of the lower platform.

6.2. Gait Parameters of 3-UPU Hexapod Walking Robot

When the upper platform moves, according to the position inverse solution Equation (6), the parameter variation of each driving motor, i.e., the equivalent link length l_i ($i = 1, 2, 3$) of the limb i can be expressed as:

$$\begin{cases} l_1 = \sqrt{(x + \Delta)^2 + y^2 + z^2} \\ l_2 = \sqrt{\left(x - \frac{1}{2}\Delta\right)^2 + \left(y - \frac{\sqrt{3}}{2}\Delta\right)^2 + z^2} \\ l_3 = \sqrt{\left(x - \frac{1}{2}\Delta\right)^2 + \left(y + \frac{\sqrt{3}}{2}\Delta\right)^2 + z^2} \end{cases} \qquad (15)$$

where (x, y, z) is coordinate of the moving platform reference point O' in the static coordinate system $O\text{-}xyz$.

When the lower platform moves, according to the inverse solution equation of the position of the mechanism, the equivalent link length l_i ($i = 1, 2, 3$) of the limb i can be obtained as:

$$\begin{cases} l_1 = \sqrt{(x - \Delta)^2 + y^2 + z^2} \\ l_2 = \sqrt{\left(x + \frac{1}{2}\Delta\right)^2 + \left(y + \frac{\sqrt{3}}{2}\Delta\right)^2 + z^2} \\ l_3 = \sqrt{\left(x + \frac{1}{2}\Delta\right)^2 + \left(y - \frac{\sqrt{3}}{2}\Delta\right)^2 + z^2} \end{cases} \qquad (16)$$

6.3. Stride Calculation of 3-UPU Hexapod Walking Robot

The hexapod walking robot is realized by alternating motion between the two platforms. Therefore, to avoid collision between the two groups of supporting legs, the maximum step size of the robot should be limited. In Figure 8a, ΔLMN and ΔEFH represent the robot's upper and lower platforms, respectively. The small squares at six vertices represent six legs. Taking the upper platform starting to move as an example, point L as the reference point, the equivalent radius of leg cross section is marked as R_T. When the upper platform moves forward along forward direction of x axis, as shown in Figure 8b, the interference between the leg of L, M and the leg of F, H, should be considered. On the premise of no collision, the expression of the maximum moving distance of the moving platform is:

$$l_{\text{max1}} = LF - 2R_T = 2(OJ - R_T) = 2(x_{G3} - R_T) \tag{17}$$

When the upper platform moves along LH direction, as shown in Figure 8c, the expression of the maximum moving distance of the moving platform is:

$$l_{\text{max2}} = LP - R_T = x_{G3} - R_T \tag{18}$$

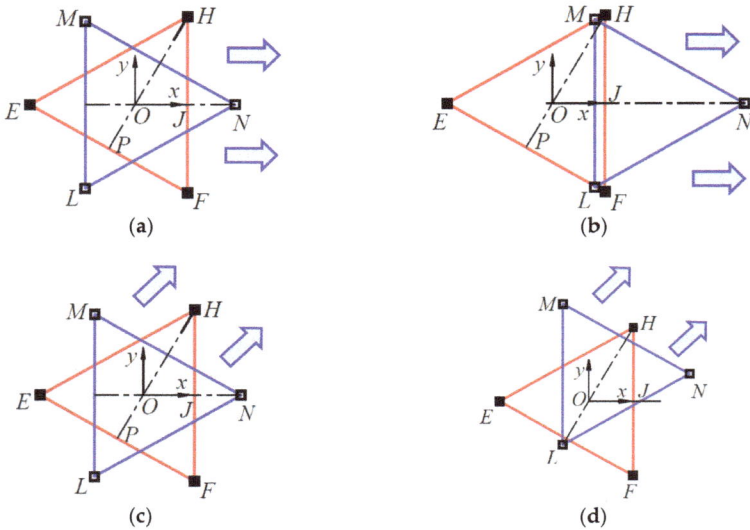

Figure 8. Motion diagram of 3-UPU hexapod walking robot. (**a**) Initial status. (**b**) Maximum moving distance of the upper platform along x axis. (**c**) Initial status. (**d**) Maximum moving distance of the upper platform along LH direction.

The allowable maximum stride of the moving platform along forward direction of x axis has to be larger than that when it moves along negative direction of x axis, as demonstrated in Figure 8. Combined with the analysis of the stability of the robot, the maximum stride of the robot is less than x_{G3} in the situation of without instability. In order to ensure the stability of the robot and avoid the interference of the legs, the maximum stride of the hexapod walking robot is set to l_{max1}. This situation occurs when the moving platform moves along the direction pointed by the triangle sharp angle of the supporting surface.

6.4. Trajectories of the Two Platforms

The trajectory of the moving platform is analyzed in related to walking on different terrains. When planning the gait of the hexapod walking robot, the trajectory curve of the moving platform can be divided into two kinds: straight line and curve. The trajectory of straight line is flexible at the same height of obstacle crossing, but at the turning point of the trajectory, the velocity and acceleration will change, which will affect the stability of the robot.

At the same obstacle height, the trajectory of the straight section is more flexible, but the mutation of the velocity and acceleration at the turning point of the trajectory occur, which affects the stability of the robot. A quadratic polynomial is thus adopted for the platform trajectory to avoid a sudden change of velocity and acceleration, whose expression is:

$$z = ax^2 + bx + c \tag{19}$$

7. Case Study

We include a case study of the analysis of the 3-UPU parallel mechanism-based walking robot. The design parameters of the 3-UPU parallel mechanism are shown in Table 2.

Table 2. Design parameters of the 3-UPU parallel mechanism.

Parameter	Value
Equivalent circumradius of the upper platform, r/mm	125
Equivalent circumradius of the lower platform, R/mm	160
Range of the equivalent link length of limb i, $l_i (i = 1, 2, 3)$/mm	$l_i \in [365, 465]$
Range of the rotation angle of the Hooke joint, $\theta_i (i = 1, 2)$/rad	$\theta_i \in [-\pi/4, \pi/4]$

Following the workspace search flow chart shown in Figure 3, the workspace of the 3-UPU parallel mechanism formed by the combination of the top and bottom surfaces and the surrounding surfaces, as shown in Figure 9. The upper and lower surfaces are formed by the limit position of the equivalent link length l_i ($i = 1, 2, 3$) of limb i. The surrounding surfaces are constrained by the limit position of the rotation angle θ_i of the Hooke joint. It was also found that the influence of the limit positions of two revolute pairs on the workspace is different.

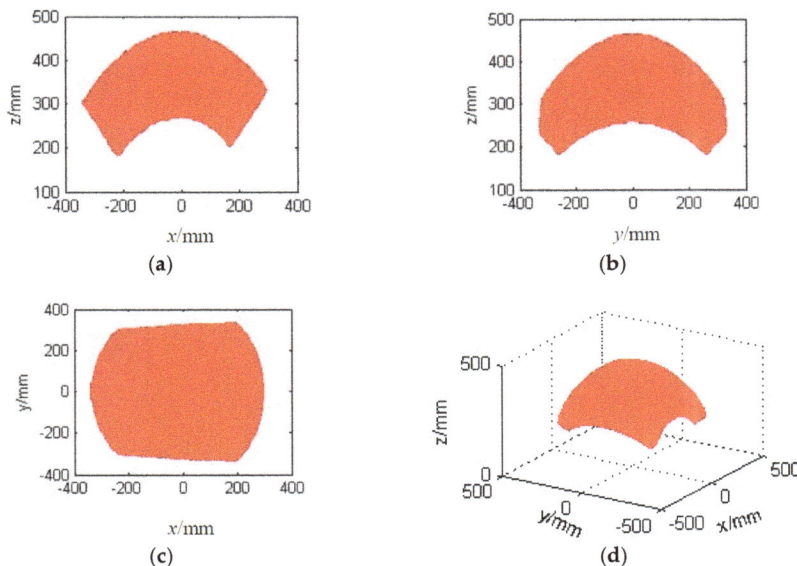

Figure 9. Workspace of the 3-UPU parallel mechanism. (a) Workspace projection in xz plane. (b) Workspace projection in yz plane. (c) Workspace projection in xy plane. (d) 3D Workspace.

In the initial assembly, the coordinates of the upper platform center in the static coordinate system is (0, 0, 363). The variation curves of the parameters of the limb lengths can be obtained when the upper platform moves using MATLAB simulation, as shown in Figure 10a,b.

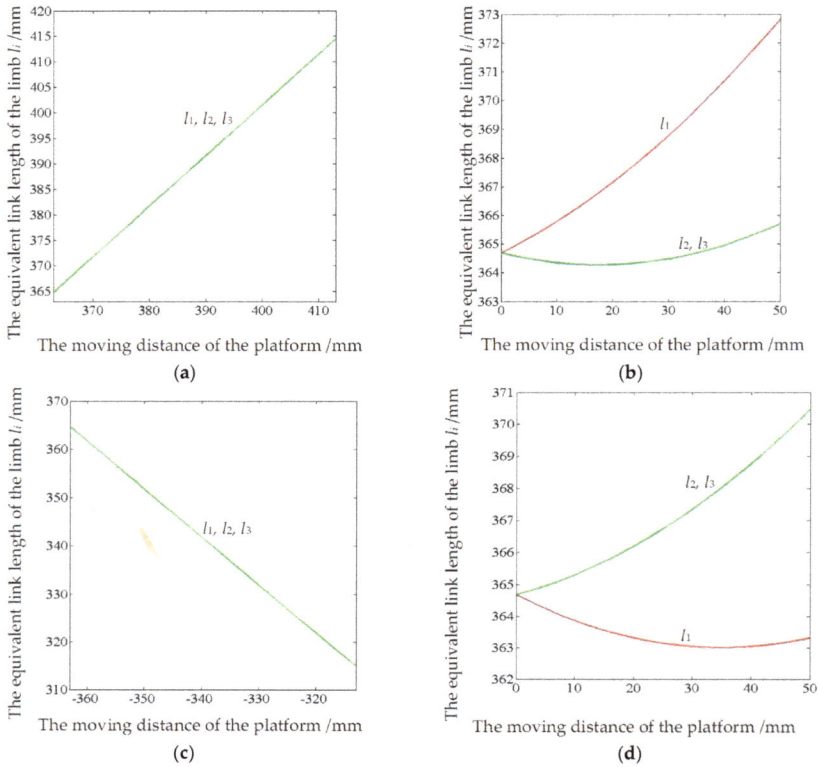

Figure 10. Variation of the limb lengths when the upper or lower platform moves. (**a**) The upper platform moves forward along *z* axis. (**b**) The upper platform moves along *x* axis. (**c**) The lower platform moves along *z* axis. (**d**) The lower platform moves along *x* axis.

When the lower platform moves, the moving and static coordinate systems exchange. The coordinate of the lower platform center in the static coordinate system is (0, 0, −363) mm. The variation curve of parameters of each driving motor can be obtained from Equation (6), as shown in Figure 10c,d. From Figure 10, the equivalent link length of each limb l_i (i = 1, 2, 3) varies smoothly when the upper platform moves, that is, the displacement of each driving motor is stable.

When the upper platform is conducted as the moving platform, the static coordinate system is established in the center of the lower platform. According to the robot's gait planning, the maximum stride l_{max1} is 160 mm, the starting point coordinate of the curve is (0, 363), and the ending point coordinate is (160, 363). According to the workspace of the 3-UPU parallel mechanism, the maximum obstacle height of the robot could cross is 80 mm, that is, the extreme point coordinate of the curve is (80, 443). Substituting the coordinates of the starting point, the end point and the extreme point into Equation (19), the trajectory equation of the reference point of the upper platform can be obtained,

$$z = -0.0125x^2 + 2x + 363 \tag{20}$$

Taking the lower platform as the moving platform, the static coordinate system is connected to the center of the upper platform. The maximum stride is 160 mm. Substituting the starting point coordinate (−60, −363), the ending point coordinate (0, −363), and the extreme point coordinate (−80,

−283) of the curve into Equation (19), the trajectory curve equation of the centroid of the reference point of the lower platform can be obtained as:

$$z = -0.0125x^2 - 2x - 363 \qquad (21)$$

Following trajectory specified by Equation (21), walking along a straight line on flat terrain with the robot was simulated in SolidWorks. A snapshot is shown in Figure 11. The variation of the stability margin is shown in Figure 12. The stability margin is calculated by $d = \min \{d_1, d_2, d_3\}$, where d_i, $i = 1, 2, 3$ are the distances from G_3 to three sides of the supporting triangle, as demonstrated in Figure 5a. As can be seen in Figure 12, the stability margin d is always larger than zero, thus static walking stability is guaranteed.

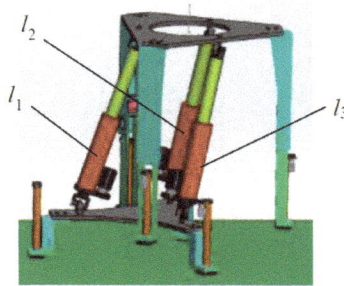

Figure 11. Walking on flat terrain of 3-UPU hexapod walking robot.

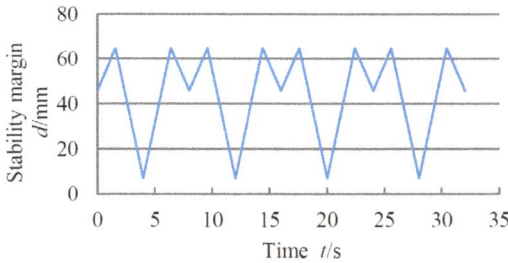

Figure 12. Variation of the stability margin when walking on flat terrain.

A simulation of walking on uneven terrain is shown in Figure 13, in which the robot walks over a ditch. The step length is adjusted to fit both the ditch width and also the stability requirement. The simulation shows that the robot can walk over the ditch stably, with the stability margin displaying in Figure 14. Compared with Figure 12 for even terrain walking, the stability margin (SM) plot in Figure 14 shows some differences. One is that the change of SM is not periodic, as the walking robot has to adjust its stride in ditch crossing. Another difference is that, in Figure 14, the SM for some short periods remains unchanged. This is because the two platforms do not move but only legs are extended to fit the terrain. It is noted that the SM in Figure 14 is always larger than zero too, which guarantees a stable static walking.

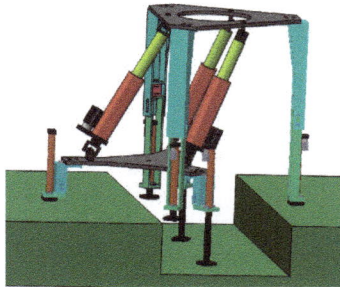

Figure 13. Crossing a ditch of 140 mm in depth and 350 mm in width by the 3-UPU hexapod walking robot.

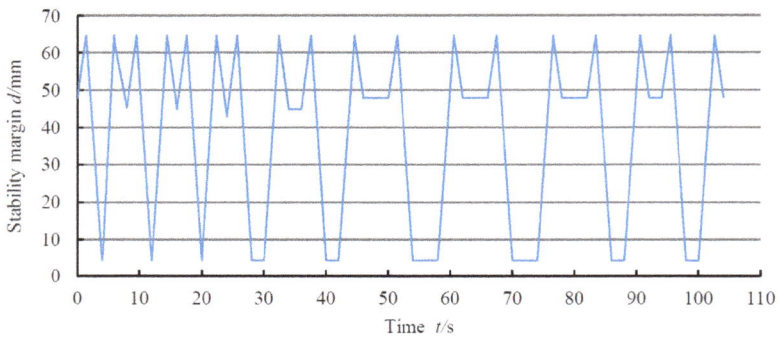

Figure 14. Variation of the stability margin when crossing ditch.

8. Discussion and Conclusions

A hexapod walking robot built with a 3-UPU parallel mechanism is introduced in this paper. The inverse kinematics solution of the mechanism platform is analyzed, the closed vector loop equation is established, and the inverse solution is obtained. The workspace of the mechanism is obtained using the constraint condition of each joint and the inverse kinematics solution.

The movement stability of the robot is analyzed using the center of gravity projection method. On the premise of maintaining the stability of the body, according to the structure parameters of the robot and without collision between the legs or between the legs and the body, the maximum transverse moving stride of the robot is 160 mm. The gait planning of the hexapod walking robot is carried out, and the variation of the equivalent link length of each limb with the movement of the robot is obtained.

The novelty of the work lies in the introducing of the 3-UPU parallel mechanism in the walking robot. While parallel mechanisms were mostly used as legs in most works [17–21], in this work, we innovatively use the parallel mechanism as both an actuator to generate walking and also a connecting body to connect two groups of three legs. The new walking robot is thus able to walk with simple gait using very few motors. Moreover, the robot has a high terrain adaptability and stability, which has been demonstrated through simulations. The robot also has other potentials, including large carrying capacity and turning ability. In future work, we will focus on the detection of the robot's center of gravity position and study further on stable walking on other uneven terrains, such as slopes and stairs. Experimental investigation with a prototype under developed is also planned.

Author Contributions: R.L. provided the ideas and wrote the paper manuscript. H.M. completed the stability and gait planning. S.B. contributed in walking analysis and major revision. Y.Y. and J.Z. tested the simulation. All authors read and approved the final manuscript.

Funding: This research was funded by the National Natural Science Foundation of China (grant number 51275486) and the Foundation of Shanxi Key Laboratory of Advanced Manufacturing Technology (grant number XJZZ201702).

Conflicts of Interest: The authors declare no conflict of interest.

References

1. Tedeschi, F.; Carbone, G. Design issues for hexapod walking robots. *Robotics* **2014**, *3*, 181–206. [CrossRef]
2. Yang, F.; Ding, X.L.; Peng, S.J. Bio-control of a modular design robot-NOROS. In Proceedings of the Third ASME/IFToMM International Conference on Reconfigurable Mechanisms and Robots, Beijing, China, 20–22 July 2015; pp. 891–900.
3. Peng, S.J.; Ding, X.L.; Yang, F.; Xu, K. Motion planning and implementation for the self-recovery of an overturned multi-legged robot. *Robotica* **2017**, *35*, 1107–1120. [CrossRef]
4. Ma, Z.R.; Guo, W.Z.; Gao, F. Analysis on obstacle negotiation ability of a new wheel-legged robot. *Mach. Des. Res.* **2015**, *31*, 6–10, 15. (In Chinese)
5. Tsai, L.W. Kinematics of a three-DOF platform with three extensible limbs. In *Recent Advances in Robot Kinematics*; Lenarcic, J., Parenti-Castelli, V., Eds.; Kluwer Academic: London, UK, 1996; pp. 401–410.
6. Gregorio, R.D. Kinematics of the 3-UPU wrist. *Mech. Mach. Theory* **2003**, *38*, 253–263. [CrossRef]
7. Ji, P.; Wu, H.T. Kinematics analysis of an offset 3-UPU translational parallel robotic manipulator. *Robot. Auton. Syst.* **2003**, *42*, 117–123. [CrossRef]
8. Han, C.; Kim, J.; Kim, J.; Park, F.C. Kinematic sensitivity analysis of the 3-UPU parallel mechanism. *Mech. Mach. Theory* **2002**, *37*, 787–798. [CrossRef]
9. Huang, Z.; Li, S.H.; Zuo, R.G. Feasible instantaneous motions and kinematic characteristics of a special 3-DOF 3-UPU parallel manipulator. *Mech. Mach. Theory* **2004**, *39*, 957–970. [CrossRef]
10. Wu, T.; Zhang, W.X.; Ding, X.L. Design and analysis of a novel parallel metamorphic mechanism. *J. Mech. Eng.* **2015**, *51*, 30–37. (In Chinese) [CrossRef]
11. Staicu, S.; Popa, C. Kinematics of the spatial 3-UPU parallel robot. *UPB Sci. Bull. Ser. D* **2013**, *75*, 9–18.
12. Lu, Y.; Shi, Y.; Hu, B. Kinematic analysis of two novel 3UPU I and 3UPU II PKMs. *Robot. Auton. Syst.* **2008**, *56*, 296–305. [CrossRef]
13. Staicu, S.; Popa, C. Dynamics of the translational 3-UPU parallel manipulator. *UPB Sci. Bull. Ser. D* **2014**, *76*, 3–12.
14. Gregorio, R.D.; Parenti-Castelli, V. Mobility analysis of the 3-UPU parallel mechanism assembled for a pure translational motion. In Proceedings of the IEEE/ASME International Conference on Advanced Intelligent Mechatronics (AIM), Atlanta, GA, USA, 19–23 September 1999; pp. 520–525.
15. Peng, B.B.; Li, Z.M.; Wu, K.; Sun, Y. Kinematic characteristics of 3-UPU parallel manipulator in singularity and its application. *Int. J. Adv. Robot. Syst.* **2011**, *8*, 54–64.
16. Zhang, D.; Wei, B. Interactions and optimizations analysis between stiffness and workspace of 3-UPU robotic mechanism. *Meas. Sci. Rev.* **2017**, *17*, 83–92. [CrossRef]
17. Miao, Z.H.; Yao, Y.A.; Kong, X.W. A rolling 3-UPU parallel mechanism. *Front. Mech. Eng.* **2013**, *8*, 340–349. [CrossRef]
18. Gu, Q.F. Kinematics of 3-UPU parallel leg mechanism used for a quadruped walking robot. In Proceedings of the 14th International Symposium on Distributed Computing and Applications for Business Engineering and Science (DCABES 2015), Guiyang, China, 18–24 August 2015; pp. 188–191.
19. Wang, M.F.; Ceccarelli, M.; Carbone, G. A feasibility study on the design and walking operation of a biped locomotor via dynamic simulation. *Front. Mech. Eng.* **2016**, *11*, 144–158. [CrossRef]
20. Sugahara, Y.; Endo, T.; Lim, H.; Takanishi, A. Design of a battery-powered multi-purpose bipedal locomotor with parallel mechanism. In Proceedings of the IEEE/RSJ International Conference on Intelligent Robots and Systems, Lausanne, Switzerland, 30 September–4 October 2002; pp. 2658–2663.
21. Wang, H.B.; Qi, Z.Y.; Hu, Z.W.; Huang, Z. Application of parallel leg mechanisms in quadruped/biped reconfigurable walking robot. *J. Mech. Eng.* **2009**, *45*, 24–30. (In Chinese) [CrossRef]

22. Hirakoso, N.; Terayama, J.; Yoshinaga, N.; Arai, T. A study on optimal formulation for multi-legged gait robot with redundant joint. In Proceedings of the International Symposium on Micro-Nanomechatronics and Human Science, Nagoya, Japan, 3–6 December 2017; pp. 1–6.

23. Sun, T.; Xiang, X.; Su, W.H.; Wu, H.; Song, Y.M. A transformable wheel-legged mobile robot: Design, analysis and experiment. *Robot. Auton. Syst.* **2017**, *98*, 30–41. [CrossRef]

24. Bai, S.; Low, K.H.; Zielinska, T. Quadruped free gait generation for straight-line and circular trajectories. *Adv. Robot.* **2000**, *13*, 513–538. [CrossRef]

25. Bai, S.; Low, K.H. Terrain evaluation and its application to path planning for walking machines. *Adv. Robot.* **2001**, *15*, 729–748. [CrossRef]

26. Bai, S.; Low, K.H.; Teo, M.Y. Path generation of walking machines in 3D terrain. In Proceedings of the IEEE International Conference on Robotics and Automation, Washington, DC, USA, 11–15 May 2002; pp. 2216–2221.

27. Gong, D.W.; Wang, P.; Zhao, S.Y.; Duan, Y. Bionic quadruped robot dynamic gait control strategy based on twenty degrees of freedom. *IEEE/CAA J. Autom. Sin.* **2018**, *5*, 382–388. [CrossRef]

28. Li, M.T.; Jiang, Z.Y.; Wang, P.F.; Sun, L.N.; Ge, S.S. Control of a quadruped robot with bionic springy legs in trotting gait. *J. Bionic Eng.* **2014**, *11*, 188–198. [CrossRef]

29. Matsuzawa, T.; Koizumi, A.; Hashimoto, K.; Sun, X.; Hamamoto, S.; Teramachi, T.; Sakai, N.; Kimura, S.; Takanishi, A. Crawling motion and foot trajectory modification control for legged robot on rough terrain. In Proceedings of the IEEE International Conference on Mechatronics and Automation, Takamatsu, Japan, 6–9 August 2017; pp. 1976–1982.

30. Winkler, A.W.; Farshidian, F.; Pardo, D.; Neunert, M.; Buchli, J. Fast trajectory optimization for legged robots using vertex-based ZMP constraints. *IEEE Robot. Autom. Lett.* **2017**, *2*, 2201–2208. [CrossRef]

31. Winkler, A.W.; Bellicoso, C.D.; Hutter, M.; Buchli, J. Gait and trajectory optimization for legged systems through phase-based end-effector parameterization. *IEEE Robot. Autom. Lett.* **2018**, *3*, 1560–1567. [CrossRef]

32. Neunert, M.; Farshidian, F.; Winkler, A.W.; Buchli, J. Trajectory optimization through contacts and automatic gait discovery for quadrupeds. *IEEE Robot. Autom. Lett.* **2017**, *2*, 1502–1509. [CrossRef]

33. Zhao, Y.; Chai, X.; Gao, F. Obstacle avoidance and motion planning scheme for a hexapod robot Octopus-III. *Robot. Auton. Syst.* **2018**, *103*, 199–212. [CrossRef]

34. Oliveira, L.F.P.; Rossini, F.L. Modeling, simulation and analysis of locomotion patterns for hexapod robots. *IEEE Lat. Am. Trans.* **2018**, *16*, 375–383. [CrossRef]

robotics

MDPI

Article

Locomotion of a Cylindrical Rolling Robot with a Shape Changing Outer Surface

Michael G. Puopolo [1],*, Jamey D. Jacob [2] and Emilio Gabino [3]

[1] Department of Mechanical Engineering, Kettering University, Flint, MI 48504, USA
[2] Department of Mechanical and Aerospace Engineering, Oklahoma State University,
    Stillwater, OK 74078, USA; jdjacob@okstate.edu
[3] Amazon Corporation, Sumner, WA 98390, USA; emilio.gabino_zarate@okstate.edu
* Correspondence: pwop@mail.com

Received: 27 July 2018; Accepted: 30 August 2018; Published: 10 September 2018

Abstract: A cylindrical rolling robot is developed that generates roll torque by changing the shape of its flexible, elliptical outer surface whenever one of four elliptical axes rotates past an inclination called trigger angle. The robot is equipped with a sensing/control system by which it measures angular position and angular velocity, and computes error with respect to a desired step angular velocity profile. When shape change is triggered, the newly assumed shape of the outer surface is determined according to the computed error. A series of trial rolls is conducted using various trigger angles, and energy consumed by the actuation motor per unit roll distance is measured. Results show that, for each of three desired velocity profiles investigated, there exists a range of trigger angles that results in relatively low energy consumption per unit roll distance, and when the robot operates within this optimal trigger angle range, it undergoes minimal actuation burdening and inadvertent braking, both of which are inherent to the mechanics of rolling robots that use shape change to generate roll torque. A mathematical model of motion is developed and applied in a simulation program that can be used to predict and further understand behavior of the robot.

Keywords: shape changing; rolling; robot; cylindrical; elliptical; velocity control; economic locomotion; actuation burden; inadvertent braking

1. Introduction

Ground based robots typically move from place to place using wheels, legs, or by changing shape in a biomimetic fashion, as with peristaltic or slithering locomotion [1–3]. Wheeled robots are the most common of these three locomotion styles because, in general, wheeled robots are highly efficient and they can move faster than other types of ground based robots [1]. A special class of wheeled robots is the rolling robot, which rolls exclusively on an outer, driven surface that entirely envelopes the system [4]. The study herein investigates locomotion of an autonomous rolling robot, developed at the Unmanned Systems Laboratory at Oklahoma State University (OSU) and pictured in Figure 1, which generates torque by means of changing its outer surface. By executing shape change at just the right time, the robot repeatedly configures itself into a forward-tilting elliptical cylinder and subsequently rolls forward under the force of its own weight. In addition to being shape changing and partially gravity powered, locomotion of the OSU Roller, as it is entitled, is categorized as dynamic [5], meaning it has a natural rocking tendency and exhibits inertial motion. In other words, if the outer surface were to suddenly stop changing shape in the middle of a roll, the robot would likely require several seconds to come to rest. Non-dynamic rollers, such as crawling rolling robots [6,7], move slowly in comparison and do not exhibit dramatic inertial effects after motion input has ceased. The OSU Roller is also an underactuated system [8], referring to how the robot exploits its own natural dynamics in order to achieve steady, rolling locomotion.

Figure 1. The OSU Roller has a mass of 0.950 kg with a perimeter of 2.095 m.

Locomotion via gravity power has previously been studied as a field of interest in biomechanics. In [9], researchers presented a mathematical model of human gait using a notional three-link machine with muscles that acted only to periodically configure the gait, leaving the machine to move entirely under the force of gravity for most of its motion. Others [10,11] investigated gaits in which gravity alone generated motion of simple link-composed machines down a slanted ramp. Authors of these studies pointed to the simplicity and efficiency of gravity powered gaits and suggested that power and control could be added using small, strategically timed energy inputs that do not interrupt the natural motion of the machine. Although the OSU Roller is not a linked walking machine like those presented in [10,11], the concept of applying strategically timed inputs during a gravity-powered gait is, in essence, how the OSU Roller works.

With the rise in popularity of modular robotics, researchers have developed shape changing rollers whereby six or more servo motor modules [2] are stacked end-to-end to form a dynamic rolling loop. In [5], researchers presented a modular loop robot composed of ten CKBot modules. On each module of the robot, there was a touch sensor for knowing when a side of the loop was in contact with the ground, at which point the loop quickly morphed into a newly configured football-like shape, now with its long axis aligned closer to the vertical, and continued rolling forward until the next side of the loop touched down. In [12], researchers constructed a modular rolling loop composed of six SuperBot modules that was programmed to deform its shape by contracting and relaxing its hexagonal shape, thereby shifting the robot center of gravity forward and causing the robot to roll. Using their hexagonal rolling loop, researchers performed an experiment that measured endurance by lapping the robot around an inside building corridor for a total distance of roughly one kilometer while voltage in the module batteries was monitored.

Modular loops, while they are highly configurable and allow roboticists to experiment with different locomotion strategies and gaits, are bulky and mechanically complicated systems [13,14]. In contrast to modular loops, researchers in [15] presented a dynamic rolling robot whose outer surface was a thin, lightweight strip of plastic that morphed by means of a singular servo motor. The robot used all onboard sensing and power to achieve velocity control relative to a desired step profile. Results of the research showed that when the robot was given a set of advantageous initial conditions, it was able to accelerate from rest and maintain constant average velocity with significant accuracy.

The robot used for the investigation herein, the OSU Roller, is a modified version of the robot presented in [15].

Although shape change has been shown to be a viable locomotion format in the aforementioned research endeavors, the subtle aspects associated with locomotion of dynamic rolling loops have previously not been unearthed. One of these aspects is how driving torque is affected by variation of shape change input timing. Along these lines, a unique characteristic of the OSU Roller is that it has the capability to measure energy consumed by its singular actuator during a trial roll, and this capability is utilized in an experiment documented herein that determines the effect of shape change actuation timing on energy consumption of the robot. Results of this experiment provide insights that can be applied to control shape changing rolling robots, including modular loop robots, more efficiently.

2. Materials and Methods

2.1. Physical Composition of the Robot

The outer surface of the rolling robot is a flat strip of polyvinyl chloride plastic put into the shape of an open, elliptical cylinder. The outer surface is firm enough to provide a steady dynamic roll surface, yet it is limber enough to be reshaped by the pull of a linear actuator positioned along the diameter of the cylinder [15]. When the actuator is not being commanded to move, its static holding force maintains a constant cylindrical shape of the outer surface. On a smooth and level floor, the robot rolls straight without leaning to the side or tipping over, and where the robot meets the floor, the outer surface bends slightly under the weight and motion of the robot. There is negligible slipping of the outer surface against the floor, as verified through high-speed video analysis. The outer surface strip measures 0.318 cm by 5.1 cm and has a perimeter of 2.095 m (Figure 1). Total mass of the robot is 0.950 kg.

Contained onboard the robot and inside the cylindrical outer surface are a microprocessor board, an inertial measurement unit (IMU) board equipped with a micro-electrical-mechanical gyroscope, two rechargeable 9 V lithium batteries that power the robot, mechanical switches that allow for continual measurement/computation of angular position of the robot, a linear actuator, and an electrical energy sensor. There is also a radio transmitter that sends data regarding robot locomotion to a receiver-equipped, external laptop computer for analysis. The microprocessor is given values of angular position and angular velocity as feedback, and brings about shape change of the outer surface via the linear actuator in order to affect velocity. These hardware components are fastened to the robot in such a way that, even as the actuator elongates and the robot changes shape, the robot center of mass remains approximately in the same location—at the central axis of the outer surface cylinder. A telescoping column orientated in a perpendicular fashion relative to the linear actuator is comprised of male and female tubes whose ends are secured to the inside of the outer surface. This telescoping column, seen in Figure 1, is not a powered actuator. Rather, it serves as an air displacement damper to limit vibration and bending of the outer surface.

In addition to the IMU board and the mechanical switches, a sensor that computes servo motor energy consumption is employed onboard the OSU Roller. At the heart of the sensor is a highly accurate current sensing chip that works in cooperation with a low ohmage resistor placed in series between the motor power supply (a regulated 5 V output from the microprocessor) and the motor load. The sensing chip receives voltage across the resistor, and in turn outputs a voltage that is representative of current that flows into the motor. The representative voltage is received at an analog-to-digital converter on a dedicated microprocessor that is separate from the main microprocessor. In a similar manner, voltage at the positive lead of the servo motor is passed to a second analog-to-digital converter on the dedicated microprocessor. During operation of the robot, a computer program running on the dedicated microprocessor samples these voltages, converts them into values of current, multiplies them for power [16], and then multiplies power by the program's sample period (200 µs) to give energy consumed by the servo motor in the span of one sample period. A running sum gives total electrical energy used by the servo motor since initiation of the program. Every tenth of a second after the computer program has begun, total electrical energy is computed, and these values are passed to the main microprocessor after a trial roll is completed.

2.2. Mathematical Model of Robot Motion

A two-dimensional model of the OSU Roller is developed in which the outer surface is an ellipse with a deflected center due to bending, as illustrated in the free body diagram in Figure 2. The ellipse rolls in one direction along a straight line in the laboratory-fixed Cartesian coordinate frame, XY, with X as the floor line. A moving rectangular coordinate frame, AB, is concentrically attached to the ellipse center and rolls with the ellipse. Axis A is coincidental with the actuator line of motion. Lengths a and b of the elliptical semi-major axes are measured along A and B, respectively. The point where the outer surface ellipse touches the floor is denominated as the touch point. The value, x_d, is the horizontal position of the touch point on X with respect to the ellipse center, and y_d is the vertical distance from X to the ellipse center. The angle defined by the positive branch of A and the vertical line passing through x_d is the roll angle of the robot, θ. (In keeping with the right-hand rule mnemonic [17], θ is negative with clockwise rotation.) As shown in Figure 2, touch angle, σ, is defined as the angle made by the A axis and the line from the ellipse center to the touch point:

$$\sigma = \arctan(\frac{x_d}{y_d}) - \theta. \tag{1}$$

Touch point location, S, is equivalent to the elliptical arc length integrated over roll angle, plus a constant value [18]:

$$S = \int \sqrt{R^2 + (R')^2} \, d\sigma + C, \tag{2}$$

with

$$R = \frac{ab}{\sqrt{a^2 \sin^2 \sigma + b^2 \cos^2 \sigma}}, \tag{3}$$

and

$$R' = \frac{ab(b^2 - a^2) \sin \sigma \cos \sigma}{\sqrt{(a^2 \sin^2 \sigma + b^2 \cos^2 \sigma)^3}}. \tag{4}$$

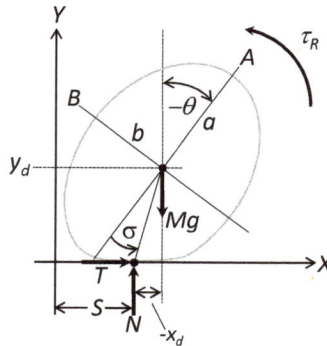

Figure 2. The robot model includes affects of bending and has three forces acting on the robot: weight, normal force and traction. The model also includes torque caused by rolling resistance.

Referring to the free body diagram in Figure 2, traction, T, and the normal force, N, act on the robot at the touch point. Weight, Mg, acts at the robot center of mass, which is located at the ellipse center. Rolling resistance torque, τ_R, opposes roll motion of the robot. Newton's second law is applied in the horizontal and vertical directions and also about the center of mass, resulting in three coupled differential equations of motion:

$$T = M(\ddot{S} - \ddot{x}_d),\tag{5}$$

$$N - Mg = M\ddot{y}_d,\tag{6}$$

$$Ty_d + Nx_d + \tau_R = \dot{H},\tag{7}$$

where $M = 0.950$ kg is total mass of the robot. $\dot{H}$ is the time derivative of the angular momentum of the robot about its center of mass, derived by modeling the robot as a collection of point masses that represent the outer surface and hardware components [15]. Equations (5)–(7) are combined into one equation of motion:

$$\ddot{\theta} = \frac{1}{\rho_1}(-\rho_2\dot{\theta} + My_d\ddot{S} - My_d\ddot{x}_d + Mx_d\ddot{y}_d + Mgx_d + \tau_R),\tag{8}$$

where ρ_1 and ρ_2 are constant coefficients of $\ddot{\theta}$ and $\dot{\theta}$, respectively, of the $\dot{H}$ term [15,19]. Variables x_d and y_d in Equation (8) are expressed as:

$$x_d = x_e + \Delta x,\tag{9}$$

$$y_d = y_e - d,\tag{10}$$

where (x_e, y_e) is location of the robot center of mass if the outer surface were a perfect ellipse, and Δx and d are changes to x_e and y_e brought on by bending of the outer surface at the floor line, as illustrated in Figure 2. By observing that X is equivalent to the tangent line of the ellipse at the touchpoint, the following expressions are derived for x_e and y_e [13]:

$$x_e = \frac{(a^2 - b^2)cos\theta sin\theta}{\sqrt{a^2cos^2\theta + b^2sin^2\theta}},\tag{11}$$

$$y_e = \sqrt{a^2cos^2\theta + b^2sin^2\theta}.\tag{12}$$

In order to quantify Δx and d, static deflection of the robot center is measured for various orientations of the robot, and the following functions are fitted in relation to an angular position parameter, $\psi = -rem(\theta, 360°)$:

$$d = 0.0093|sin\psi|,\tag{13}$$

$$\Delta x = \begin{cases} 0.005|sin2\psi|, & \text{if } 0 \leq \psi < 180°, \\ -0.005|sin2\psi|, & \text{if } 180° \leq \psi < 360°, \end{cases}\tag{14}$$

Rolling resistance torque of the outer surface is modeled as $\tau_R = k_R d\dot{\theta}^2$, where k_R is a positive value that changes with $\dot{\theta}$. This model rightly predicts that $\tau_R = 0$ when $d = 0$ and when $\dot{\theta} = 0$. In addition, because d and $\dot{\theta}^2$ are never negative during trial rolls, the model rightly predicts that τ_R always acts in the clockwise direction, resisting rotation of the robot whenever it moves and bends. The power of two on angular velocity causes the model to capture sharp decelerations that are observed only when the robot moves with relatively high angular velocity. In order to determine appropriate values of k_R, three trial rolls of the robot are performed at three different steady state angular velocities. In the first of these rolls, the robot is controlled to reach -2.0 rad/s and continue, on average, at this velocity for eight seconds. While the roll is in progress, angular velocity of the rolling robot is sampled using the onboard gyroscope and recorded as a function of time. Afterwards, the simulation program (described subsequently in this section) is run using the same initial conditions as the trial roll. Values of k_R are iteratively used in the program until the simulated response best matches that of the actual robot. Matching is done by comparing total distance rolled and plots of angular velocity at steady state. The value of k_R used in the best match simulation is taken as the appropriate model value: 1.6 N/s^2. This process is repeated for -2.2 rad/s, and the appropriate value of k_R is found to be 2.1 N/s^2. For -2.4 rad/s, the appropriate value of k_R is found to be 1.9 N/s^2.

2.3. Robot Velocity Control System

Shape change actuation is triggered when axis *A* or axis *B* leans into the roll and passes a certain inclination. When this triggering happens, the robot control system immediately causes the linear actuator to either extend or contract, and the shape of the robot is changed [15]. Upon completion of shape change, the linear actuator remains at the newly changed length until actuation is again triggered. Consider the illustration in Figure 3, in which the rolling robot is shown rolling to the right when shape change actuation is triggered by axis *B*. The control system responds by changing linear actuator length along axis *A*, and, consequently, outer surface eccentricity changes as the robot continues to roll to the right. Roughly a quarter-turn after actuation commences, two scenarios are shown in Figure 3. If *a* has been made longer as *A* leans into the roll, the robot undergoes an induced torque imbalance about its center of mass that pushes the robot forward and increases speed, as illustrated by the lower ellipse in Figure 3. On the other hand, if *a* is changed so that the ellipse is now a circle, there is no resulting torque imbalance due to offset, and average speed is not increased.

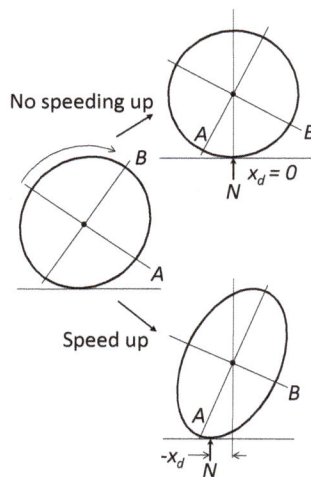

Figure 3. The moment, Nx_d, causes a torque imbalance about the robot center of mass, affecting roll dynamics of the robot.

Shape change actuation is triggered using the concept of tilt. Tilt angle of the robot is

$$\gamma = -rem(\theta, 90°) \tag{15}$$

and can be thought of as forward inclination of the robot measured either by *A* or *B* relative to the vertical. Tilt angle is used by the control system to perform two actions. First, a special measurement of robot angular velocity ($\dot{\theta}$) is performed by the onboard gyroscope when γ newly becomes greater than or equal to a set angle, θ_m, which is usually 35° for the trial rolls conducted in the experiment documented herein in Section 2.5.2. This special measurement of angular velocity is referred to as ω_m. The second action is actuation triggering, which occurs when γ newly becomes greater than or equal to the set trigger angle, θ_t. In general, as desired speed of the robot increases, θ_t must be decreased, or else there is not enough time for the robot to change shape before the next trigger moment. For the experiment documented herein in Section 2.5.2, θ_t varies from 25° to 65°.

Upon actuation triggering, the microprocessor prepares for shape change by performing several computations to determine target length of the linear actuator. The first of these computations is that of error:

$$Error = k_s \omega_d - \omega_m, \tag{16}$$

where ω_d is the desired angular velocity, and k_s is a positive constant that is slightly greater than unity. Scaling the desired velocity by k_s is necessary because even when the robot is rolling with nearly the desired velocity (that is, when $\omega_m \approx \omega_d$), the robot must speed up to compensate for the natural slowing of rolling resistance in order to stay close to the desired angular velocity by the next trigger moment. With *Error* computed in this manner, it is in turn used to compute an intermediary value of the target length for the semi-major axis that will be next to lean into the roll [15]. That length is given by

$$l_t = R_c - k_f \times Error, \tag{17}$$

where R_c is equal to 0.3335 m, the radius of the outer surface while in the circular configuration, and k_f is a positive control constant. If *Error* is negative (the robot is rolling to the right too slowly), the control system prepares to elongate the semi-major axis relative to the circular configuration. If *Error* is positive (the robot is rolling to the right too quickly), k_f is temporarily set to zero until the next trigger moment. In this latter case, the robot assumes the circular configuration and is slowed by rolling resistance. Final target length of the axis is computed according to a saturation operation that ensures the ellipse stays within the physical bounds of the system [15]:

$$L_t = \begin{cases} s_{min} & \text{if } l_t \leq s_{min}, \\ l_t & \text{if } s_{min} < l_t < s_{max}, \\ s_{max} & \text{if } l_t \geq s_{max}, \end{cases} \tag{18}$$

where s_{min} and s_{max} are the smallest and largest allowable values of the outer surface semi-major axis due to rotation limits of the actuator servo motor, and they are equal to 0.319 m and 0.349 m. Since the linear actuator lies along A, if A is the next axis to lean into the roll, the control system causes the linear actuator to elongate or shorten so that a is equal to L_t. If B is the next axis to lean into the roll, the control system changes a so that b becomes equal to L_t. Time duration of the actuation depends on the orientation of the linear actuator and the magnitude of shape change; when the robot rolls with steady state velocity, time duration of actuation is roughly one-third of a second.

2.4. Simulation Program

A Matlab/Simulink computer program (version R2017a, MathWorks, Natick, MA, USA) is developed to solve Equation (8) numerically for θ as a function of roll time. The program code is a loop structure that uses initial conditions for θ and a to solve the equation for $\ddot{\theta}$ at the first pass of the loop. After this first pass, the program "loops back" values that are needed in order to compute $\ddot{\theta}$. As part of this process, the program checks γ at each pass to determine if actuation has been triggered, and if it has, a, b, $\dot{a}$ and $\dot{b}$ are newly computed; $\dot{a}$ and $\dot{b}$ are modeled as trapezoidal velocity profiles that, when integrated, give a and b as functions of roll time. Updated values of a, b, $\dot{a}$, $\dot{b}$, ρ_1, and ρ_2 are then looped back, along with values for θ and $\dot{\theta}$ obtained through numerical integration, which the program performs using Simulink's Runge–Kutta [20] solver, and $\ddot{\theta}$ is newly computed. Due to difficulty in finding exact expressions for some of the derivatives on the right side of Equation (8), a numerical derivative algorithm is employed to find the following values: $\dot{y}_d$, $\ddot{y}_d$, $\dot{x}_d$, $\ddot{x}_d$, and $\ddot{S}$. Once $\ddot{\theta}$ is newly computed, it is integrated yet again, and the process is repeated until the simulated roll is completed. At each pass of the loop, motion parameters of the robot, including $\dot{\theta}$, are logged and can be plotted as a function of roll time after the program has terminated.

2.5. Experimental Procedure

2.5.1. Control Program

The robot moves under control of a program that is run onboard the microprocessor. The control program is written in C++ programming language, and is essentially an invocation of one primary function [21] named loop that repeats every 10 ms. Prior to loop, the control program has two precursor sections of code that each run only once when the program is executed. The first is a variable definition section, where several variables are set by the user to form a combination of control values and initial conditions that define the trial roll. The variables that can be set are: θ_m, θ_t, k_f, k_s, ω_d and $a(0)$. After the variable definition section, the program performs various initialization tasks, including enablement of microprocessor ports, sensors, communication lines, and moving the actuator to the set value of $a(0)$. After these precursor sections have run, the loop function is finally invoked, but before loop is executed for the first time, the control program sends the energy sensor a digital high voltage signal, causing the sensor to begin the process of measuring current and voltage into the servo motor to compute energy consumed. Roll time of a trial roll begins when this signal is sent, and an LED on the microprocessor board is illuminated as indication. When roll time goes beyond 15 s, the control program electronically detaches the servo motor and enters an idle state until it is manually turned off when the robot is retrieved.

As described in a previous section, position switches are placed on the outer surface of the robot in a manner such that voltage in the switch circuit goes low when one of the position switches is activated, and activation occurs whenever A is in the vertical position, i.e., $-180°$, $-360°$, $-540°$, etc. In the loop function, the control program checks the digital line connected to the switch circuit to see if voltage has gone low since the last cycle of loop. If it has, a flag is momentarily turned on in the program, and a count is tallied of how many times the flag has been turned on since the trial roll began. If loop sees the flag presently on, roll angle of the robot is computed from the flag count. If loop sees the flag presently off, θ is computed in the sensor program by numerically integrating the sampled angular velocity, according to the trapezoidal rule [20], starting from the last angle measured by the switch circuit.

With θ computed at every cycle of loop, the control program uses θ to compute γ and checks for initiation of angular velocity measurement and an actuation trigger. If angular velocity measurement has been initiated, the current reading of the gyroscope is saved as ω_m. If actuation has been triggered, the control program computes *Error* and L_t, and then initiates shape change actuation of the robot. Regardless of triggering, at every cycle of loop, the control program computes the predicted values of a, b, and their time-derivatives. Roll distance, S, is also computed at every cycle of loop using an implementation of Equation (2).

The rolling robot possesses a radio transmitter, by which the microprocessor communicates with a receiver-equipped laptop computer located near the roll track. Every 100 ms (or every tenth cycle of loop), the control program transmits several locomotion-related parameters to the laptop, where the parameters are printed as a line of comma separated values to a serial monitor window. The transmitted/printed parameters are: θ, ω (which is the same as $\dot{\theta}$), S, a and total energy. In addition, just before entering an idle state, the control program reads the total energy array from the energy sensor and transmits energy values to the laptop computer for every 100 ms of roll time.

In Figures 4 and 5, angular velocity during two example trial rolls of the robot are plotted versus roll time. In Figure 4, the robot displays stable locomotion, and, in Figure 5, the robot displays unstable locomotion. Stability, as used herein, refers to two conditions being met relative to controlled angular velocity of the robot. The first condition is the robot must speed-up quickly from rest, so that rise time is no more than 6 s. If rise time is longer than 6 s, there is not enough length on the roll track to facilitate a sufficient span of steady state motion for analysis. The second condition is that after robot velocity has risen to ω_d, a running average, ω_r, of its measured angular velocity must thereafter remain close to the desired constant velocity. Some reasonable variation of ω_r within several degrees per second is permissible, but large and fluctuating dips in angular velocity or a steady increase of ω_r over time are considered unstable response characteristics.

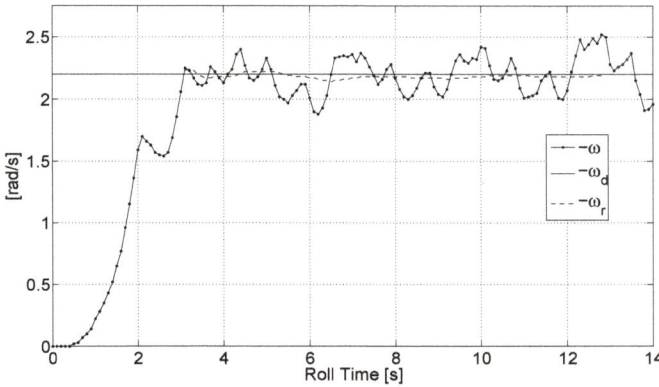

Figure 4. An example of stable locomotion of the robot for $\omega_d = -2.2$ rad/s. Angular velocity rises quickly to the desired level and remains close to it thereafter.

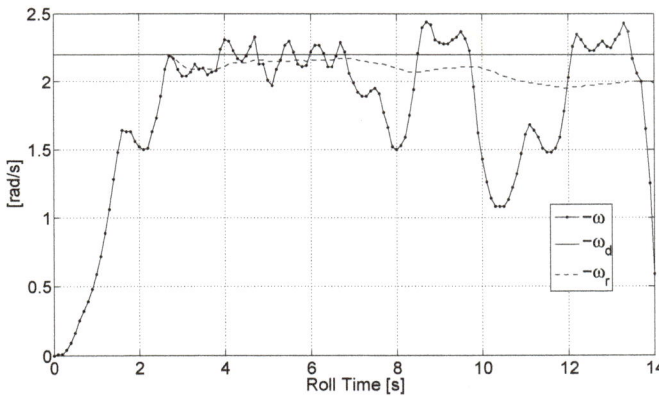

Figure 5. An example of unstable locomotion for the robot for $\omega_d = -2.2$ rad/s. Average angular velocity dips far below the desired level during the latter half of the trial roll.

2.5.2. Trial Rolls

Experimental trial rolls are conducted and provide information used to solve the following three optimization problems:

Find θ_t that minimizes ϱ for

1. $\omega_d = -2.0$ rad/s with $\theta_t \in \{35°, 45°, 55°, 65°\}$
2. $\omega_d = -2.2$ rad/s with $\theta_t \in \{25°, 35°, 45°, 55°\}$
3. $\omega_d = -2.4$ rad/s with $\theta_t \in \{25°, 35°, 45°\}$

where ϱ is energy consumed per unit roll distance of the robot servo motor, and it is calculated according to a method described in Section 2.5.3. Values of θ_t that are suitable for stable locomotion for the various angular velocities have been identified before the experiment is conducted in order to quantify the constraints on θ_t for Problems 1–3. Combinations of k_f, k_s and θ_t that result in minimization of ϱ for the various velocities have also been identified previously and are used in the trial rolls conducted for Problems 1–3.

The experiment starts by fully charging the batteries and allowing the servo motor to remain unpowered for at least three hours to reach the ambient temperature of the laboratory. Next, columns of the linear actuator are cleaned and lubricated, and various parameters in the control program are set. For Problem 1, the desired velocity, ω_d, is set to -2.0 rad/s, θ_t is set to $35°$, and θ_m is set to $35°$. Ten preliminary trial rolls of the robot are then conducted in order to warm-up the servo motor. All trial rolls in the experiment are conducted on the same roll track (Figure 6) with a consistent set of initial conditions: $\theta(0) \approx -10°$, $\omega(0) = 0$, and $a(0) = 0.349$ m.

Figure 6. Roll track for the robot is a flat and level section of laminate flooring installed on concrete.

A random number generator is employed to shuffle the order of the trial rolls for Problem 1 relative to viable values of θ_t. Shuffling is done to avoid any would-be biasing caused by rising temperature of the servo motor. Then, using the order of θ_t from the shuffled sequence, 40 trial rolls are conducted in order to collect information to address Problem 1. The first of these trial rolls is conducted with θ_t set to the first entry in the sequence, and information from the trial roll is saved. Afterward, θ_t is set to the second entry in the sequence, and a second trial roll is conducted, timed to start 45 s after the first roll ends, and information from the trial roll is saved. By the time the last value in the sequence is used, ten trial rolls are conducted for each value of θ_t. After trial rolls for Problem 1 are finished, columns of the linear actuator are newly cleaned/lubricated, the batteries are fully recharged, and the motor is allowed to cool. The roll sequence is reshuffled, and ω_d is changed to -2.2 rad/s in order to address Problem 2. For Problem 3, ω_d is changed to -2.4 rad/s. For all problems, the value of θ_m is set to $25°$ when $\theta_t = 25$; otherwise, it is set to $35°$.

Upon completing trial rolls for Problems 1–3, saved roll parameters are used to calculate ϱ for each roll, resulting in ten-member ϱ populations that each correspond to a value of θ_t for a given problem. Uncertainty in the calculation of ϱ, a quantity that is based largely on propagation of measurement errors associated with total energy, is also calculated for each trial roll. From there, central tendencies of the ϱ populations for a given problem are compared using a computerized implementation of the Monte Carlo method [22], in which the populations are repeatedly perturbed according to the largest uncertainty and subjected to the median test [23]. Results of the comparison are used to rank populations for a given problem with regard to central tendency. As it turns out, due to uncertainty, it is impossible to confidently establish one population as having the least positive central tendency for any given problem; rather, for each problem, at least two populations are determined to have central tendencies that are equivalent and yet significantly less positive than all others. Corresponding values of θ_t for these populations are deemed superior because they result, on average, in lowest energy consumption per unit roll distance of the robot. Further details of the aforementioned Monte Carlo process and how superiority of θ_t is determined can be found in [24].

2.5.3. Energy Per Unit Roll Distance, ϱ

As the robot rolls in a controlled manner, various resistive agents perform non-conservative work that retards motion, causing the robot to move slower than it would if these agents were not present. The primary resistive agents are: servo motor inefficiency, friction in columns of the linear actuator, and rolling resistance torque. For a given roll distance, ΔS, of the robot, the sum of non-conservative work performed by these agents is a negative value denoted herein as W_{NC}, and non-conservative work performed per unit distance rolled is defined as

$$\varrho \equiv -\frac{W_{NC}}{\Delta S}. \tag{19}$$

The negative sign used in the definition is meant to ensure that ϱ is always positive, a convention adopted merely for convenience in reporting results. When comparing two trial rolls at a given velocity, the roll that registers a lower value of ϱ is less-burdened and is thus a more economical roll. In addition to being an indicator of burden, ϱ can be thought of as the amount of electrical energy required to move the robot a roll distance of one meter. In the experiment documented herein, values of ϱ for trial rolls of the robot are compared in order to determine which values of θ_t result in the most economical locomotion in terms of energy consumption.

After an experimental trial roll is completed, the robot control system reports locomotion-related parameters collected during the trial roll at every tenth of a second to a laptop computer. Reported parameters are saved in a digital spreadsheet as successive rows containing values of θ, ω, S, a, and total energy consumed by the servo motor. In order to calculate ϱ for each trial roll, two roll time points are chosen from the spreadsheet. Point 1 is chosen from points at the beginning of the trial roll after the robot has achieved steady state velocity and when the robot is in, or is very close to, the circular configuration (with $a = 0.3335$ m). Point 2 is chosen from points at the end of the roll when the robot is close to being in the circular configuration. Even though selection of Points 1 and 2 vary from trial roll to trial roll, they almost always envelop about nine seconds of steady state velocity of the robot.

After Points 1 and 2 have been chosen, the work-energy equation [17] is applied, by which the following conservation equation is derived:

$$E_m + W_{NC} = \Delta K + \Delta U, \tag{20}$$

where ΔK and ΔU are changes in kinetic and potential energy, respectively, of the robot from Point 1 to Point 2, and E_m is total energy supplied to the servo motor between Points 1 and 2. W_{NC} is the non-conservative work performed between Points 1 and 2. Change in kinetic energy, ΔK, in Equation (20) is calculated according to the assumption that the robot is a rigid body:

$$\Delta K = \frac{1}{2}Mv_2^2 + \frac{1}{2}I\omega_2^2 - \frac{1}{2}Mv_1^2 - \frac{1}{2}I\omega_1^2, \tag{21}$$

where I is mass moment of inertia of the robot about its center of mass in the circular configuration (0.071 kg·m^2), and v and ω are subscripted to signify linear and angular velocities at Points 1 and 2. Because the robot is in, or nearly in, the circular configuration at Points 1 and 2, height of robot center of mass is equal to $h - d$, where h is radius of the unloaded outer surface cylinder in the circular configuration. Treating $h - d$ as the radius of the rolling robot, the arc length formula is applied in order to calculate velocity of robot center of mass at Points 1 and 2, resulting in: $v_1 = (h - d_1)\omega_1$ and $v_2 = (h - d_2)\omega_2$, where d_1 and d_2 are changes in height of robot center of mass due to bending at Points 1 and 2. Substituting for v_1 and v_2, Equation (21) becomes

$$\Delta K = \frac{M}{2}[(h - d_2)^2\omega_2^2 - (h - d_1)^2\omega_1^2] + \frac{I}{2}(\omega_2^2 - \omega_1^2). \tag{22}$$

Potential energy of the robot at any point is equal to gravity potential of the robot center of mass plus potential stored in deflection of the flexible outer surface. For the purpose of calculating these values, an imaginary reference datum is placed at the surface of the roll track, and the outer surface is treated as having variable stiffness, $k_s = Mg/d$. Accordingly, potential energy of the robot at Point 1 is

$$U_1 = Mg(h - d_1/2) \tag{23}$$

and potential energy at Point 2 is

$$U_2 = Mg(h - d_2/2). \tag{24}$$

The change in potential energy between Points 1 and 2 is

$$\Delta U = U_2 - U_1 = \frac{Mg}{2}(d_1 - d_2). \tag{25}$$

As defined previously, ϱ is equal to the negative of non-conservative work, W_{NC}, performed between Points 1 and 2 divided by the distance, $\Delta S = S_2 - S_1$, traveled between Points 1 and 2. Combining this definition with Equations (20), (22) and (25) gives

$$\varrho = -\frac{Mg}{2\Delta S}(d_1 - d_2) - \frac{M}{2\Delta S}[(h - d_2)^2 \omega_2^2 - (h - d_1)^2 \omega_1^2] - \frac{I}{2\Delta S}(\omega_2^2 - \omega_1^2) + \frac{E_m}{\Delta S}. \tag{26}$$

3. Results

Figure 7 is a tabulated summary of results from the trial roll experiment, in which superior values of θ_t are identified for the various values of ω_d. In addition, the simulation program is configured to perform three rolls of the robot. Each of the simulated rolls has a set of initial conditions and control constants that are identical to a trial roll conducted in the experiment. Roll 1 is from Problem 1 with $\theta_t = 35°$, $\theta(0) = -14°$, and $\omega_d = -2.0$ rad/s. Roll 2 is from Problem 2 with $\theta_t = 45°$, $\theta(0) = -14°$, and $\omega_d = -2.2$ rad/s. Roll 3 is from Problem 3 with $\theta_t = 35°$, $\theta(0) = -7°$, and $\omega_d = -2.4$ rad/s. By design, the rolls have desired angular velocities and trigger angles whose collective values span the ranges of these parameters tested in the experiment. The simulation program is configured to output locomotion parameters, including ω, at 10 millisecond intervals of roll time for the simulated trial rolls. Output values of angular velocity from the simulated rolls are gathered and plotted along with angular velocity from actual trial rolls, and the resulting comparison plots for Rolls 1–3 are included in Figures 8–10.

	25°	35°	45°	55°	65°
-2.0 rad/s	▓	✕	●	●	✕
-2.2 rad/s	●	●	●	✕	▓
-2.4 rad/s	●	●	✕	▓	▓

Figure 7. Cells with a dot signify values of θ_t that are superior in terms of energy economy of robot locomotion. Gray cells signify values of θ_t that cause instability, and cells with a cross-out signify inferior values of θ_t.

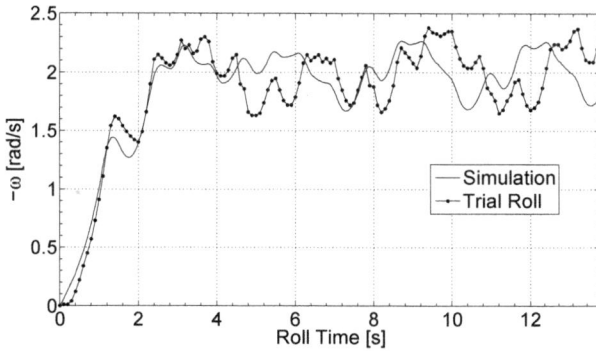

Figure 8. Robot angular velocity from simulation and from trial roll are plotted versus roll time for Roll 1, in which $\omega_d = -2.0$ rad/s.

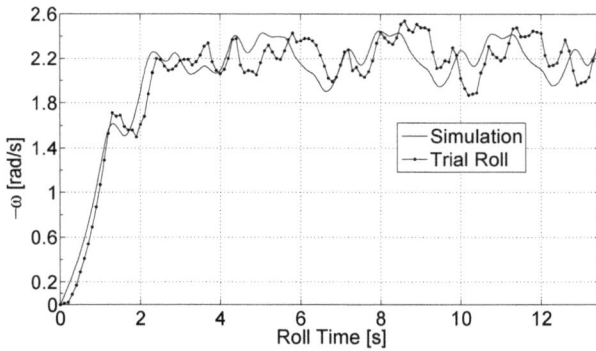

Figure 9. Robot angular velocity from simulation and from trial roll are plotted versus roll time for Roll 2, in which $\omega_d = -2.2$ rad/s.

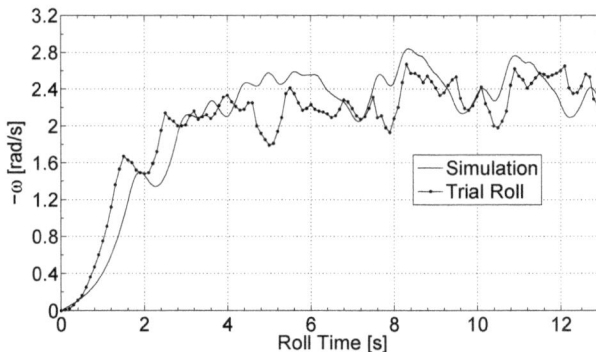

Figure 10. Robot angular velocity from simulation and from trial roll are plotted versus roll time for Roll 3, in which $\omega_d = -2.4$ rad/s.

4. Discussion

The most interesting takeaway from the research documented herein is that, for each ω_d tested in the experiment, there exists a range of two or three values of θ_t that are determined to be superior in terms of energy economy of robot locomotion. When θ_t is decreased relative to this range, so that actuation is triggered with less tilt of the triggering axis with respect to the vertical, energy economy of robot locomotion is observed to decrease; and when θ_t is increased relative to this range, so that actuation is triggered with more tilt of the triggering axis, energy economy of robot locomotion is again observed to decrease. There is therefore an optimal range of θ_t with regard to energy economy for each ω_d tested in the experiment. Furthermore, results of the experiment reveal that as desired angular velocity of the robot is changed, the optimal range of θ_t changes as well. In general, this optimal range shifts downward (i.e., less tilt of the triggering axis) as velocity magnitude of the robot is increased from 2.0 rad/s to 2.4 rad/s.

In order to understand the existence of this range of θ_t and why it shifts with robot speed, recall that the robot control system works by repeatedly morphing outer surface shape in order to change location of the normal force relative to the robot center of mass. In this way, the control system ensures that the normal force is most often located to the left of the center of mass during stable locomotion of the robot, causing input torque to be applied in a clockwise sense to drive the robot along X in the positive direction and to follow, on average, a desired angular velocity. However, upon careful scrutiny of controlled motion of the robot, it is apparent that the control system, which actuates shape change at most four times per revolution of the robot, has mixed consequences—that is, the control system speeds up the robot, but it sometimes slows it down as well.

Imagine the roll scenario illustrated in Figure 11 where roll instances are arranged chronologically from left to right. In this scenario, θ_t is set below the optimal range for $\omega_d = -2.0$ rad/s, meaning that actuation occurs "early" in the γ cycle when $\gamma < 45°$. At the first instance in the scenario, A has just swept past the trigger angle, and shape change actuation has begun. The moment arm, x_d, is negative because oblongness of the outer surface, combined with the slight tilt of the robot, place the touchpoint to the left of the robot center of mass. With a negative moment arm, Nx_d is clockwise in the first instance, and the robot speeds up as a result. Because shape change actuation occurs early in this scenario, the outer surface of the robot actuates to the circular configuration, as in the second instance in Figure 11, before A crosses the horizontal orientation. In the circular configuration, Nx_d is zero. At instance three, the robot has become oblong about B, and A still has not crossed the horizontal, so that x_d has become positive and Nx_d is now counterclockwise, and the robot briefly undergoes inadvertent braking. At instance four, A has rotated into the horizontal, at which point x_d and Nx_d inevitably become zero again, and the scenario subsequently repeats with A and B in switched positions by the fifth instance.

A symbolic representation is introduced to characterize the input torque pattern on the robot in this early actuation scenario, wherein a plus sign is used to represent counterclockwise torque on the robot, and a negative sign is used to represent clockwise torque. The "0" digit is used to represent zero torque on the robot, and the expression, $\gamma = 0°$, is used to represent the instance when A or B is oriented horizontally, when gamma and Nx_d are both zero. Starting at the first instance and using this symbolic representation, the pattern displayed by the early actuation scenario is: $-$, 0, +, $\gamma = 0°$, $-$, 0, +, $\gamma = 0°$, $-$, etc. A marked feature of this actuation pattern is the sequence: +, $\gamma = 0°$, $-$; that is to say, when $\gamma = 0°$, torque on the robot is changing from counterclockwise to clockwise.

For the purpose of learning if the robot exhibits the early actuation pattern during actual locomotion, four individual trial rolls from Problem 1, one each viable value of θ_t, are chosen and investigated. The values of ϱ for the chosen rolls are consistent with average values of the corresponding ϱ populations. The four trial rolls are referred as Roll A, Roll B, Roll C and Roll D, respectively corresponding to $\theta_t = 35°$, $45°$, $55°$, and $65°$. Equation (14) is used with reported roll data to compute x_d at every tenth of a second for each roll. In Figure 12, computed values of x_d are plotted versus roll time for Roll A. Dashed, vertical lines in the graph represent approximate moments during the roll when $\gamma = 0°$. A plot of angular velocity versus roll time for Roll A is also included in Figure 13.

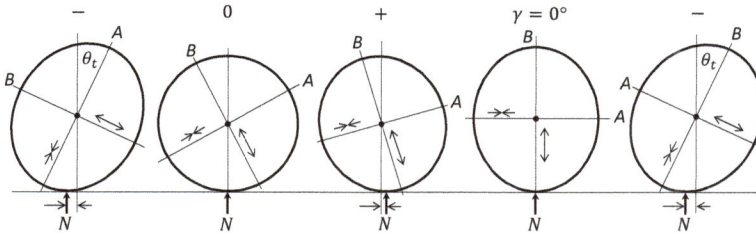

Figure 11. When the robot exhibits the early actuation pattern, inadvertent braking occurs before the robot has reached the $\gamma = 0°$ orientation.

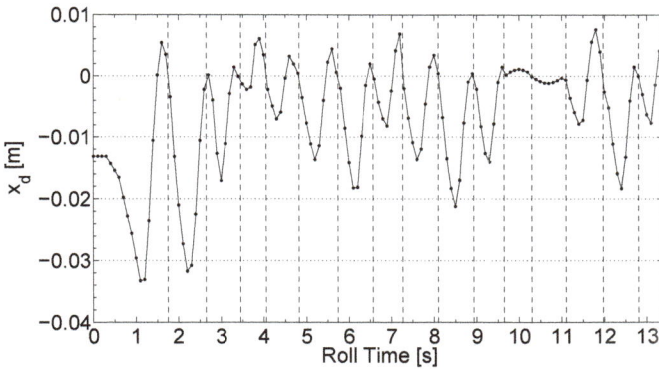

Figure 12. For Roll A, the robot exhibits the early actuation pattern, in which inadvertent braking (when x_d is positive) causes the roll to have relatively low energy economy.

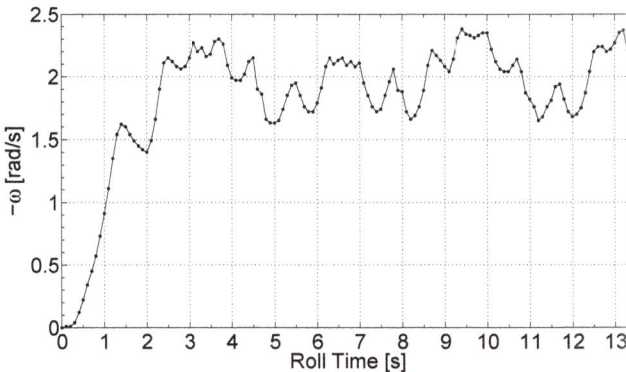

Figure 13. Angular velocity of the robot for Roll A, a roll with relatively low energy economy. At times, the robot slows itself down when it is already going too slow relative to the desired velocity of -2.0 rad/s.

The plot of x_d in Figure 12 reveals that x_d is frequently positive, which means the rolling robot experiences repeated inadvertent braking during steady state. These times are relatively short-lived, and the magnitude of x_d during inadvertent braking is generally smaller than when x_d is negative, and this result means driving torque supplies more energy to the robot than does inadvertent braking, which is expected, since the purpose and demonstrated outcome of the control system is to maintain forward motion of the robot. The actuation pattern for Roll A is clearly that of early actuation. The graph in Figure 12 shows 13 times at which $\gamma = 0°$, and at ten or more of these times, the x_d pattern is: $+$, $\gamma = 0°$, $-$, which is that of early actuation. For Roll A, θ_t is set to 35°, which is below the optimal range of θ_t identified for $\omega = -2.0$ rad/s, so it is not surprising that the early actuation pattern would manifest here.

The early actuation pattern manifested during Roll A sheds light on why there is a lower limit on optimality of θ_t in Problem 1. When θ_t is set too low (actuation triggering is too early), the robot is prone to inadvertent braking, potentially at every γ cycle. Notice from Figure 12 that, when braking occurs, it is not always beneficial. That is to say, the robot often works against itself with considerable energy during the roll, slowing itself down when it is already going too slow. A good example occurs at roughly 5.5 s. At this time, the robot is already rotating too slowly (see Figure 13) relative to the desired velocity magnitude when inadvertent braking comes on at 5.8 s and slows down the robot even more. Consequently, the linear actuator is subsequently forced by the control system to move at a large magnitude to create driving torque in order to speed up the robot and maintain the desired velocity. A similar thing happens between 11 and 12 s. With so much ill-timed braking, it is understandable why Roll A with $\theta_t = 35°$ is not optimal in terms of energy economy.

Similar investigations into Rolls B, C and D reveal that there is nothing about the respective plots of x_d that points to why robot locomotion during Rolls B and C are relatively economical compared to Roll D, as is established by the results of the experiment (Figure 7). This absence of contrast leads one to believe there is a factor separate from actuation patterning that affects energy economy of rolls in Problem 1. In an effort to identify this factor, reported information from Roll D is scrutinized, and it is noticed that high energy consumption rates during Roll D correspond to periods of actuation when A is the trigger axis. With this correspondence in mind, the average rate of energy consumption by the servo motor is calculated for actuation when A is the trigger axis, and it is found to be higher, on average, than when B is the trigger axis or during periods when the actuator is holding a constant. The reason higher trigger angle values are associated with greater actuation burden is that deformation of the outer surface is greater during such actuation moves, resulting in outwardly directed end effects that pull on the actuator and resist contraction. In contrast, when orientation of A during actuation is closer to the vertical, there is less bending of the telescoping columns and hence less friction; furthermore, gravity actually helps the actuator contract when it is orientated in an upright orientation. The effect of orientation of A on actuation burden is illustrated in Figure 14.

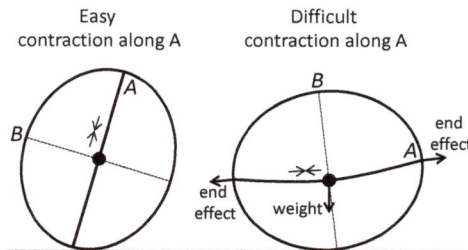

Figure 14. When A is oriented nearly vertically, gravity makes contraction "easy" for the linear actuator. When oriented horizontally, weight and outwardly directed end effects on the linear actuator make contraction "difficult".

With these insights, energy economy results from Problem 1 can now be understood as stemming from a combination of inadvertent braking and actuation burden. To summarize, at one extreme when θ_t is set low at 35°, angular velocity of the robot, which is -2.0 rad/s on average, is small enough in magnitude so the actuating outer surface often attains the circular configuration before $\gamma = 0°$. In other words, the robot exhibits the early actuation pattern, as illustrated in Figure 11, in which it frequently undergoes inadvertent braking and works against itself. For this reason, robot locomotion has relatively low energy economy for $\theta_t = 35°$. At the other extreme, when θ_t is set high at 65°, actuation becomes increasingly burdensome due to the effects of gravity and bending of the outer surface, thereby nullifying gains in economy that might otherwise be had. In between 35° and 65°, there is an optimal range where θ_t is high enough for the robot to largely avoid the averse early actuation pattern, yet low enough so that actuation burden does not greatly hinder the system. The same explanation given here for energy economy in Problem 1 applies to trial rolls in Problems 2 and 3, but with the difference that at higher rotation speeds, actuation burden is shifted downward on θ_t relative to Problem 1, so that $\theta_t = 55°$ and 45° are rendered least economical for Problems 2 and 3, respectively, while the lower trigger angles tested for each of these problems are most economical.

5. Conclusions

A rolling robot has been developed that generates torque by changing shape of its elliptical outer surface, which is flexible and can be morphed to retain oblongness about one of two notional, elliptical axes that are fixed to the robot and roll with it. The robot has been equipped with a sensing/control system by which it measures its angular position and angular velocity, computes error with respect to a desired step velocity profile, and changes the shape of its outer surface accordingly. Shape change actuation occurs four times per revolution, whenever an elliptical axis rotates past a predetermined trigger angle. The robot has demonstrated stability during roll tests, in the sense that it was able to quickly reach a constant desired angular velocity and remain close to it thereafter.

A series of trial rolls of the robot were performed using various trigger angles, while energy consumed by the servo motor was measured and used to calculate energy economy for each roll. Results of this experiment showed that, depending on the velocity of the robot, there exists a range of trigger angle values that are superior in terms of energy economy. This region of optimality on trigger angle generally shifts towards the vertical as desired angular velocity is increased. In search for an explanation, it was found that economical trial rolls featured a synchronicity of actuation timing and angular velocity, wherein the robot avoided agents of inefficiency that slowed the robot or burdened the servo motor. At higher magnitude angular velocities of the robot, actuations triggered at sufficiently small tilt angles preserved the synchronicity.

A mathematical model was developed for the robot that included bending of the outer surface and rolling resistance torque. Based on the model, a computer program was developed that simulated locomotion of the robot and was used to plot various motion parameters such as angular velocity and roll distance versus roll time. The program was configured to perform three simulation runs corresponding to actual trial rolls of the robot. Angular velocity from the simulations was compared to measured values from actual trial rolls, and the model was found to be significantly accurate.

In future research efforts, the authors would like to investigate the possibility of making an extraction of the OSU Roller with an outer surface that is bigger or smaller in circumference than the OSU Roller. Having developed a mathematical model of the robot, the effects on performance of such changes in size could easily be investigated in theory. In addition, perhaps future shape changing rolling robots could be constructed of adaptive materials such as Nitinol, pneumatic muscles, or a thermally tunable foam that acts as frame, wheel and actuator all-in-one. Imagining such extractions, which are presumably not too far off in the future, highlights how shape changing rollers have the potential to be simple, ultralightweight, efficient robots for use in various applications, including space exploration [25].

Author Contributions: E.G. contributed to the design of the robot energy sensor and revised the paper; J.D.J. contributed to the design of the experiment, interpreted data, and performed substantial revisions to the paper; M.G.P. designed and manufactured the robot, programmed the robot, contributed to the design of the experiment, interpreted data, performed the experiment, developed the mathematical model, and wrote the paper.

Funding: This research was funded in part by the Unmanned Systems Research Institute at Oklahoma State University.

Acknowledgments: The authors would like to acknowledge helpful discussions with Russ Rhinehart and donations of materials for the robot by Slate Rehm, Darby Rehm, and Chanten Rehm.

Conflicts of Interest: The authors declare no conflict of interest.

References

1. McKerrow, P.J. *Introduction to Robotics*; Chapter Components and Subsystems; Addison Wesley Publishing Company: Sydney, Australia, 1991; p. 57.
2. Bogue, R. Shape Changing Self-Reconfiguring Robots. *Ind. Robot Int. J.* **2015**, *42*, 290–295. [CrossRef]
3. Watanabe, W.; Sato, T.; Ishiguro, A. A Fully Decentralized Control of a Serpentine Robot Based on the Discrepancy between Body, Brain and Environment. In Proceedings of the 2009 IEEE/RSJ International Conference on Intelligent Robots and Systems, St. Louis, MO, USA, 10–15 October 2009; pp. 2421–2426.
4. Armour, R.; Vincent, F. Rolling in Nature and Robotics: A Review. *J. Bionic Eng.* **2006**, *3*, 195–208. [CrossRef]
5. Sastra, J.; Chitta, S.; Yim, M. Dynamic Rolling for a Modular Loop Robot. *Int. J. Robot. Res.* **2009**, *28*, 758–773. [CrossRef]
6. Sugiyama, Y.; Hirai, S. Crawling and Jumping by a Deformable Robot. *Int. J. Robot. Res.* **2006**, *25*, 603–620. [CrossRef]
7. Matsuda, T.; Murata, S. Stiffness Control—Locomotion of Closed Link Robot with Mechanical Stiffness. In Proceedings of the 2006 IEEE International Conference on Robotics and Automation, Orlando, FL, USA, 15–19 May 2006; pp. 1491–1498.
8. Tedrake, R. Underactuated Robotics: Algorithms for Walking, Running, Swimming, Flying and Manipulation (Course Notes for MIT 6.832). 2016. Available online: http://underactuated.mit.edu (accessed on 8 March 2017).
9. Mochon, S.; McMahon, T.A. Ballistic Walking. *J. Biomech.* **1980**, *13*, 49–57. [CrossRef]
10. McGeer, T. Passive Dynamic Walking. *Int. J. Rob. Res.* **1990**, *9*, 62–82. [CrossRef]
11. Tavakoli, A. Gravity powered locomotion and active control of three mass system. In Proceedings of the ASME 2010 Dynamic Systems and Control Conference, Cambridge, MA, USA, 12–15 September 2010; pp. 141–148.
12. Chiu, H.C.; Rubenstein, M.; Shen, W.M. Multifunctional SuperBot with Rolling Track Configuration. In Proceedings of the 2007 IEEE/RSJ International Conference on Intelligent Robots and Systems, San Diego, CA, USA, 29 October–2 November 2007.
13. Mellinger, D.; Kumar, V.; Yim, M. Control of Locomotion with with Shape-Changing Wheels. In Proceedings of the 2009 IEEE International Conference on Robotics and Automation, Kobe, Japan, 12–17 May 2009; pp. 1750–1755.
14. Melo, K.; Velasco, A.; Parra, C. Motion Analysis of an Ellipsoidal Kinematic Closed Chain. In Proceedings of the 2011 IEEE IX Latin American Robotics Symposium and IEEE Colombian Conference on Automatic Control, Bogota, Colombia, 1–4 October 2011.
15. Puopolo, M.G.; Jacob, J.D. Velocity control of a cylindrical rolling robot by shape changing. *Adv. Robot.* **2016**, *30*, 1484–1494. [CrossRef]
16. Chitode, J.; Bakshi, U. *Power Devices and Machines*; Technical Publications Pune: Maharashtra, India, 2009; pp. 7.20–7.23.
17. Serway, R.A. *Physics for Scientist and Engineers*, 3rd ed.; Saunders College Publishing: Philadelphia, PA, USA, 1990; Chapters 7, 27, pp. 152–173, 740–760.
18. Chandrupatla, T.R.; Osler, T.J. The Perimeter of an Ellipse. *Math. Sci.* **2010**, *35*, 122–131.
19. Hibbeler, R. *Dynamics*, 12th ed.; Chapter Kinetics of a Particle: Impulse and Momentum; Prentice Hall: Upper Saddle River, NJ, USA, 2010; pp. 262–297.
20. Pal, S. *Numerical Methods*; Oxford University Press: Oxford, UK, 2013; Chapters 14, 15, pp. 439–440, 522–538.
21. Savitch, W. *Absolute C++*; Chapter C++ Basics; Pearson: Boston, MA, USA, 2013; p. 4.

22. Chou, Y. *Statistical Analysis*; Chapter Queing Theory and Monte Carlo Simulation; Holt, Rinehart & Winston of Canada: Toronto, ON, Canada, 1969; pp. 720–730.
23. Pett, M.A. *Nonparametric Statistics for Health Care Research*; Chapter Addressing Differences Among Groups; Sage Publications: Thousand Oaks, CA, USA, 1997; pp. 204–211.
24. Puopolo, M.G. Locomotion of a Cylindrical Rolling Robot with a Shape Changing Outer Surface. Ph.D. Thesis, Oklahoma State University, St. Walter, OK, USA, 2017.
25. Antol, J.; Calhoun, P. *Low Cost Mars Surface Exploration: The Mars Tumbleweed*; Technical Report NASA/TM-2003-212411; National Aeronautics and Space Administration: Washington, DC, USA, 2003.

robotics

MDPI

Article

Design and Experiments of a Novel Humanoid Robot with Parallel Architectures[†]

Matteo Russo [1,*], Daniele Cafolla [2] and Marco Ceccarelli [1]

[1] Laboratory of Robotics and Mechatronics, University of Cassino and Southern Latium,
    Via Di Biasio 43, 03043 Cassino FR, Italy; ceccarelli@unicas.it
[2] IRCCS Neuromed, Via Atinense 18, 86077 Pozzilli IS, Italy; bioingegneria@neuromed.it
* Correspondence: matteo.russo@unicas.it; Tel.: +39-0776-299-3395
† This paper is an extended version of our paper published in Russo, M.; Cafolla, Daniele; Ceccarelli, M.
    Development of LARMbot 2, A Novel Humanoid Robot with Parallel Architectures. In Proceedings of the
    2018 4th IFToMM Symposium on Mechanism Design for Robotics, Udine, Italy, 11–13 September 2018;
    pp. 17–24.

Received: 1 November 2018; Accepted: 2 December 2018; Published: 4 December 2018

Abstract: In this paper, the mechanical design of the LARMbot 2, a low-cost user-oriented humanoid robot was presented. LARMbot 2 is characterized by parallel architectures for both the torso and legs. The proposed design was presented with the kinematics of its main parts—legs, torso, arms—and then compared to its previous version, which was characterized by a different leg mechanism, to highlight the advantages of the latest design. A prototype was then presented, with constructive details of its subsystems and its technical specifications. To characterize the performance of the proposed robot, experimental results were presented for both the walking and weight-lifting operations.

Keywords: humanoid robots; parallel mechanisms; cable-driven robots; robotic legs

---

1. Introduction

The design of humanoid robots has been one of the key challenges of robotics in the last decades, and the most successful solutions are all based on serial architectures, since the focus on humanoids is usually on control and artificial intelligence. Therefore, research on alternative mechanical designs is limited, especially if it is based on parallel architectures, despite the architectures better mechanical performance. Research on humanoid robots started fifty years ago with the development of the first humanoid robot, the WABOT-1 of Waseda University [1], and it has been a hot topic ever since. In the last ten years, several successful humanoid designs were released by both academia and industry. They are currently used as an open-source platform for research on navigation, interaction, and learning. An example is the robot NAO by Aldebaran Robotics (now SoftBank Robotics), launched in 2008, that is nowadays the standard platform for several robotics competitions, such as the RoboCup Standard Platform League [2]. Another example is the iCub robot, conceived as the platform for research on cognitive development [3]. Some other examples of humanoid robots are the WALK-MAN, a rescue robot developed for unstructured environments [4]; Pepper, manufactured by SoftBank Robotics, which is focused on human–robot interaction [5]; WABIAN-2, one of the most recent humanoids at Waseda University [1,6]; Ami, a humanoid robot for applications in domotics [7]; REEM-B by PAL-Robotics, designed to help humans in daily tasks [8]; and ARMAR, another collaborative robot for home automatization [9,10]. From a mechanical point of view, all these robots are characterized by serial architecture, since the large workspace and mobility of the 5R (where R stands for Revolute joint), 6R, 7R, and 8R kinematic chains that are used for arms and legs allow them to imitate human motion and dexterity. However, the payload of these structures is rather small (for example, NAO can lift approximately 0.15 kg per arm) and most of the structures are characterized by poor dynamic performance. For these reasons, a parallel architecture can be used to improve accuracy, payload, and dynamics. A full humanoid robot with parallel architecture, the LARMbot, was developed at the LARM laboratory of the University of Cassino and Southern Latium in 2015, as documented in References [11,12]. LARMbot was conceived as a service robot for autonomous walking and manipulation tasks. It is based on two parallel subsystems, one for the legs and one for the trunk. While the trunk design, that is shown in Reference [13], is characterized by a good kinematic and dynamic performance [14], the leg design in Reference [15] has several issues that prevent the functioning of the entire system. The 3UPU (where U stands for Universal joint and P for Prismatic joint) parallel architecture of each leg shows constraint singularities in its workspace that hinder the motion. Furthermore, the workspace of the leg mechanism is small when compared to its size, with a step length that is equal to 0.3 times the leg height, significantly smaller than the human step (which is approximately 0.8 times the human leg). For this reason, this paper presents the LARMbot 2, with a novel leg mechanism. The proposed leg design was based on a 3UPR architecture that was characterized by no singularity of any kind in its reachable workspace, as shown in Reference [16], and optimized to have a larger step than the previous design (approximately 0.8 times the leg height) [17]. In this paper, the implementation of a novel leg mechanism in the LARMbot humanoid was described. First, the mechanical design of LARMbot 2 was introduced with a description of its main subsystems and degrees of freedom. Then, a prototype was produced to validate the novel leg design, which was tested for constrained walking and weight-lifting operations.

2. Design of LARMbot 2

LARMbot 2 is characterized by three main mechanical subsystems, namely locomotion, manipulation, and torso. The locomotion subsystem is composed of two identical leg units. The kinematic scheme of the leg is shown in Figure 1a and its kinematics and dynamics are discussed and analyzed in detail in Reference [16,17]. Each leg unit is characterized by a hybrid structure with a 3UPR lower-mobility parallel mechanism that connects the hip to the ankle and that is actuated by three linear actuators in the links. An additional rotational motor is placed on the ankle for an additional degree of freedom of the foot platform, to achieve balance during walking operations by better reacting to disturbances on the frontal plane (XZ plane in Figure 2). Each leg unit has four degrees of freedom (three translational and a rotational one), for a total of eight degrees of freedom for the locomotion subsystem of the humanoid. With respect to the previous 3UPU leg design, the novel leg design is characterized by a larger workspace with no singular configurations, as detailed in Reference [16], owing to its special joint design. As shown in the kinematic scheme in Figure 1b, this joint design is characterized by three revolute joints around axes that converge at the center of the platform.

The manipulation subsystem is composed of two arm units, which are based on the kinematic scheme shown in Figure 1c. The upper arm is a 3R serial chain with two rotational degrees of freedom in the shoulder and an additional revolute joint in the elbow. The hand is a cable-driven mechanism, characterized by a 3R structure for each finger apart from the thumb, which is a 2R serial chain. Each revolute joint has a limited rotational motion of $\pi/2$ rad controlled by a small torsional spring placed between the consecutive phalanxes or palm and the first phalanx. The springs keep each finger open in a straight position, unless a cable that runs into it is pulled. Each finger is driven by a different cable, and all the cables are attached to a pulley driven by a servomotor located in the wrist. Thus, each arm unit is characterized by an underactuated mechanism with 17 rotational degrees of freedom (3 for the upper arm, and 14 for the hand), where the motion is regulated by 4 servomotors and 14 torsional springs.

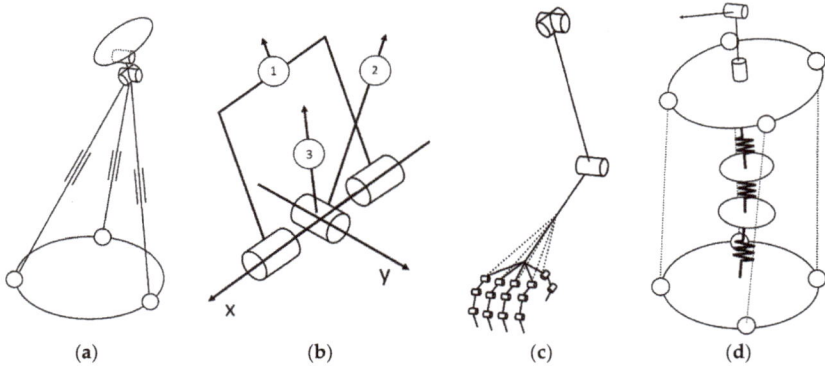

Figure 1. Kinematic scheme of LARMbot 2: (**a**) Leg; (**b**) Leg 3R joint design; (**c**) Arm; (**d**) Torso.

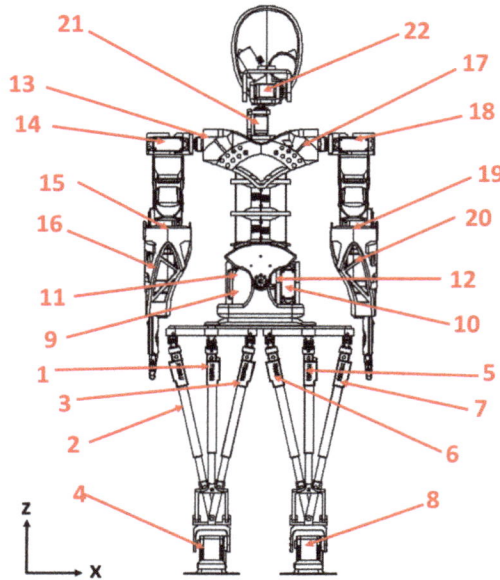

Figure 2. Actuators of LARMbot 2.

The torso is based on the CAUTO design presented in Reference [13], which is a cable-driven, underactuated hybrid manipulator based on the kinematic scheme in Figure 1d. The lower part of the torso consists of an underactuated serial chain, composed of rigid bodies and elastic joints (E) with combined spherical and translational mobility alternating in a 3E chain. Four cables with varying length (assimilable to an SPS chain) are connected in parallel to the 3E chain to control the relative position of the upper torso platform with respect to the hip platform. The entire lower-torso architecture can be described as a 4SPS-(3E) parallel mechanism with 4 degrees of freedom, which are actuated by the four motors that regulate the length of each cable. The upper torso has two additional rotational degrees of freedom in the neck, for a total of 6 degrees of freedom. As outlined in Reference [18], the motion of the torso can be used to enhance and support walking balance. Therefore, CAUTO's mobility compensates for the missing degrees of freedom in each leg, allowing the LARMbot 2 to achieve balance and react to disturbances parallel to the sagittal plane (YZ plane, with reference to Figure 2).

In Table 1, the main features of the LARMbot 2 are summarized. The entire humanoid is driven by 22 motors. Sixteen motors are rotational motors, while six are linear actuators. The position of the motors is shown in Figure 2, with additional details on the type and location of motors shown in Table 2.

Table 1. Modules of LARMbot 2.

Module	Abbr.	W (mm)	D (mm)	H (mm)	Mass (kg)	Actuators
Left Leg	LL	160	150	400	0.5	3 leg, 1 ankle
Right Leg	RL	160	150	400	0.5	3 leg, 1 ankle
Torso	TO	200	150	300	1.2	4 cables
Left Arm	LA	60	60	360	0.5	3 arm, 1 hand
Right Arm	RA	60	60	360	0.5	3 arm, 1 hand
Head	HD	95	150	150	0.4	2 neck
LARMbot	-	320	150	850	3.6	22

Table 2. Degrees of Freedom of LARMbot 2 as in Figure 2.

DoF	Location	Description	Force/Torque	DoF	Location	Description	Force/Torque
q_1	LL	Linear actuator B	36 N (4.5 mm/s)	q_{12}	TO	Cable servomotor FR	1.08 Nm
q_2	LL	Linear actuator L	36 N (4.5 mm/s)	q_{13}	LA	Shoulder motor 1	1.08 Nm
q_3	LL	Linear actuator R	36 N (4.5 mm/s)	q_{14}	LA	Shoulder motor 2	1.08 Nm
q_4	LL	Ankle servomotor	1.08 Nm	q_{15}	LA	Elbow motor	1.08 Nm
q_5	RL	Linear actuator B	36 N (4.5 mm/s)	q_{16}	LA	Hand motor	1.08 Nm
q_6	RL	Linear actuator L	36N (4.5 mm/s)	q_{17}	RA	Shoulder motor 1	1.08 Nm
q_7	RL	Linear actuator R	36N (4.5 mm/s)	q_{18}	RA	Shoulder motor 2	1.08 Nm
q_8	RL	Ankle servomotor	1.08 Nm	q_{19}	RA	Elbow motor	1.08 Nm
q_9	TO	Cable servomotor BL	1.08 Nm	q_{20}	RA	Hand motor	1.08 Nm
q_{10}	TO	Cable servomotor BR	1.08 Nm	q_{21}	HE	Neck motor 1	1.08 Nm
q_{11}	TO	Cable servomotor FL	1.08 Nm	q_{22}	HE	Neck motor 2	1.08 Nm

3. Prototype Construction

LARMbot 2 was conceived as a low-cost humanoid robot. Therefore, it was designed to be manufactured through 3D printing [19], controlled by commercial boards [20,21], and driven by commercial servomotors and linear actuators [22,23]. The cost of all the components for the final prototype was lower than 2000€. A CAD model of the humanoid robot can be seen in Figure 3, whilst a prototype is shown Figure 4.

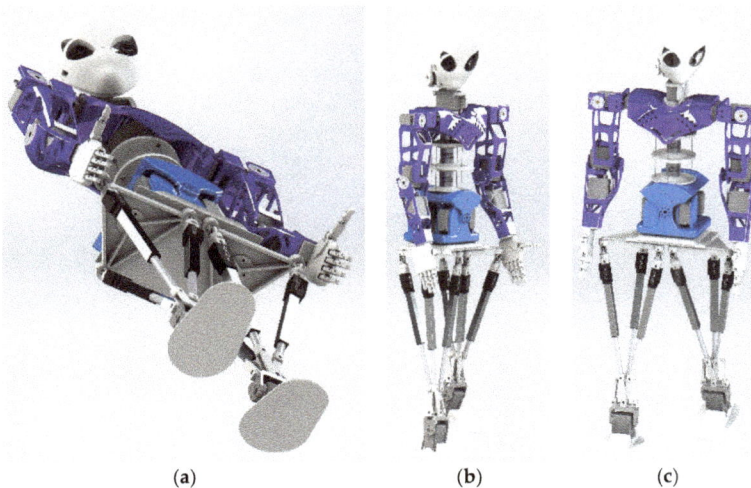

Figure 3. CAD model of LARMbot 2: (**a**) Lower view; (**b**) Side view; (**c**) Front view.

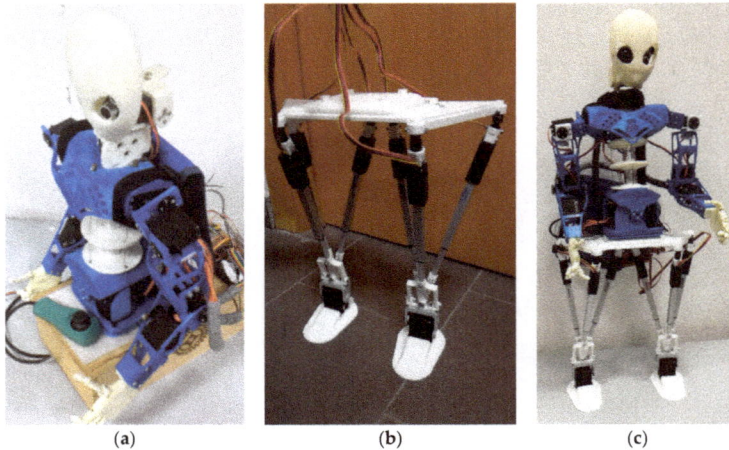

Figure 4. Prototype of LARMbot 2: (**a**) Torso and arms; (**b**) Legs; (**c**) Full assembly.

The locomotion module of the prototype is less than 320 mm wide, 150 mm deep, and 400 mm high. Its weight is 1.05 kg, considering both the mechanical structure and electronics. The upper body (torso, head, and manipulation module) is 320 mm wide, 150 mm deep, and 450 mm high, and its weight is equal to 2.60 kg. Thus, the entire prototype is 850 mm tall and has a total weight of approximately 3.70 kg (approximately 2.00 kg for the motors and the 3D printed frame, and 1.70 kg for the control boards, sensors, and battery), making the entire system compact and lightweight. Its payload capability for manipulation is 0.85 kg, limited by the serial structure of the arm, whilst the parallel architecture of the torso and legs allows for a theoretical payload up to 3.00 kg. The payloads were evaluated for the peak efficiency point of the linear servomotors and at the estimated torque at operating speed for the rotational servomotors.

The hand of LARMbot 2 is one of the most challenging components to design and manufacture, with its five-finger cable-driven structure, as shown in the model in Figure 5. Its size is 24 × 110 × 90 mm. Each finger is actuated by a single cable, which runs through a φ2.83 mm guide running through the palm. The cables are characterized by a 0.23 mm diameter and are Dyneema cables in gel spun polyethylene, which is a synthetic fiber designed for traction strength. All the cables are attached to a single servomotor in the wrist, which controls the closure of the hand by pulling all the cables together. When the cables are released, the opening movement is performed by torsional springs that are enclosed within each finger and the palm, and within the consecutive phalanxes of the same finger. An exploded view of a single finger with the torsional springs is reported in Figure 6b. This mechanism design allows the hand to adapt to different shapes of grasped objects, while still being able to lift objects weighting up to 1.00 kg.

Figure 5. LARMbot 2 hand design: (**a**) Hand model; (**b**) Exploded view of a finger.

The prototype was validated with both dynamical simulations and Finite Element Analysis (FEA). A detailed characterization of the upper body was reported in Reference [13]. A dynamic simulation of a step was performed to identify the most critical load configurations on the leg. In those configurations, the proposed design was validated with a FEM analysis that was characterized by the load of the upper body with the maximum payload applied to the upper platform of a single leg. The FE simulation assumed all the 3D-printed components as ABS bodies with a linear elastic and isotropic behavior (tensile strength equal to 3×10^7 N/m^2), with a fixture constraint which locks the foot to the ground and a load equal to the weight of the upper body plus a 36 N external payload applied normally to the upper platform of the leg. The system was meshed with 4 Jacobian points, maximum element size of 6.67 mm and minimum element size of 0.33 mm. The results in Figure 6 highlight the capability of the proposed design to withstand the maximum load. In particular, Figure 6a shows the load distribution on the leg structure, with a maximum stress of 5.55×10^3 N/m^2, and Figure 6b reports the corresponding factor of safety (FOS) on the commercial components. The critical component was the nut of the universal joint, which still showed a FOS of 1.7 in the most critical load configuration, whilst the actuators and the 3D printed components could easily withstand the stress.

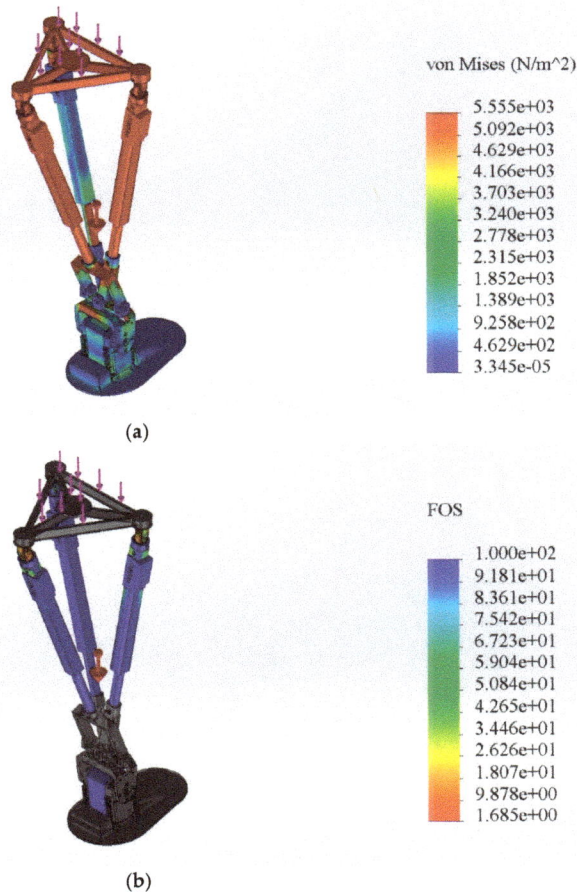

(a)

(b)

Figure 6. Leg FEA: (**a**) Von Mises equivalent stress; (**b**) Factor of Safety distribution.

4. Control and Sensing

LARMbot 2 is equipped with several onboard sensors, which range from Inertial Measurement Units (IMUs) to current sensors and cameras. In particular, its head is equipped with an Inertial Measurement Unit (IMU) that is encased in the back of the neck. The right eye is a Wi-Fi mini-camera that can acquire and record audio and video, with a resolution of 640 × 480 px at 30 fps. Its view angle is equal to 60°, and it can transmit information up to 15 m. The left eye stores an ultrasonic distance sensor that is able to detect the distance of the closest object in front of it from 20 mm up to 3 m. An example of the head sensors in function is reported in Figure 7. The figure shows the data acquired by the sensors through a software interface from a computer, where it is possible to see the signal that is acquired from the ultrasonic distance sensor on the left side. The upper window on the right is the real-time acquisition from the Wi-Fi mini-camera of the right eye, while the lower window on the right is an external camera that shows the head with the surrounding environment. In Figure 7a the head is facing the empty room, and no obstacle is detected by the ultrasonic distance sensor. In Figure 7b, the head is facing a cardboard box instead, and the distance sensor detects the obstacle at a distance of 130 mm, as reported by the interface, which also gives a graphical representation of the field of view (green lines are open field of view, red lines are no field of view).

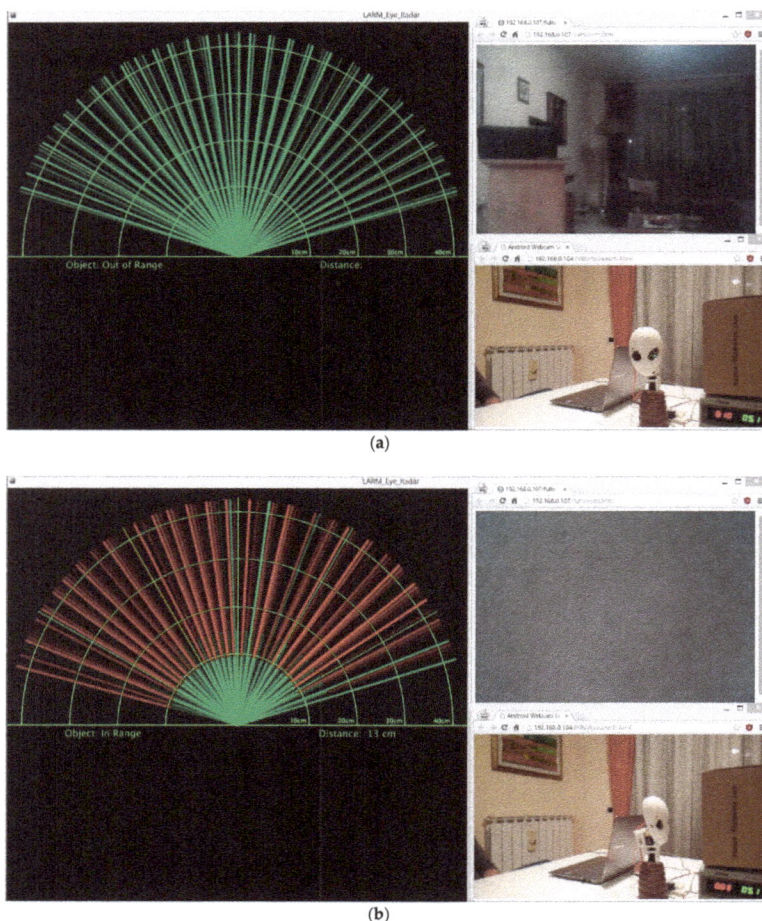

(a)

(b)

Figure 7. Test of the head sensors: (**a**) No obstacle in range; (**b**) Obstacle in range.

The control system of the novel locomotion module was based on Arduino boards [20], with a single Arduino Nano board with a PID control for both legs, passing through an Adafruit PWM Servo Driver board [21] to control the servomotors [22,23] with a velocity-based PID controller. An Inertial Measuring Unit (IMU) [24] was placed on the moving platform of the parallel mechanism for each leg to measure the angular displacement and linear acceleration of the prototype. A current sensor was used to measure the power consumption, since the power supply was set to a constant 6.8 V voltage. Another Arduino Nano board controls all the sensors. A computer was used both as remote the control for the locomotor and to acquire and store the data from the sensors. The control system is summarized by the conceptual scheme in Figure 8, and motion control was based on the closed-form expression of kinematics as presented in Reference [16].

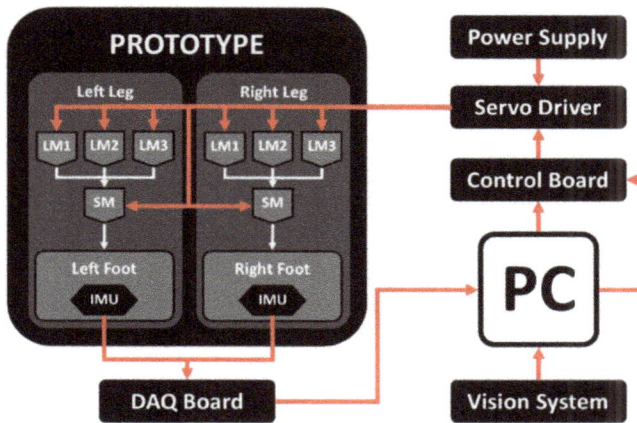

Figure 8. A conceptual scheme for the control of the biped locomotor module.

5. Experimental Validation

To characterize the step size, walking, and weight lifting performance of the novel locomotion module, two different experiments were performed: the first one was a constrained walking operation of the locomotion module, whilst the second one was a weight-lifting operation of the entire humanoid. The walking operation was characterized by a rectangular step of length equal to 200 mm and a height equal to 30 mm. The walking step trajectory was optimized by using the Output Transmission Index as the criterion [16]. The Transmission Index is a virtual coefficient that represents the virtual power delivered by a unit transmission wrench on the corresponding unit output twist of the target body. Thus, it characterizes the force transmission performance of a robot. The chosen trajectory was contained in a region of the workspace with an Output Transmission Index greater than 0.55. The input of the linear actuators for the chosen path were calculated through Inverse Kinematics and are shown in Figure 9a. In Figure 9b eight different snapshots of the test are shown as starting position, left pre-swing, left swing, double support with left forward, right mid-swing, right swing, double support with right forward, and left mid-swing, respectively. Owing to the leg design, the steps are wide even for a short motion of the sliding actuated links.

The accelerometers and gyroscopes of the IMUs were used to measure both the angular displacement and the linear acceleration of the feet of the prototype. Since during the walking operation the path of the right leg is equal to the one of the left leg but with a different phase, it is possible to use the results of a single leg to characterize the motion. Several tests consisting of five steps each were performed, with similar results. The results shown in this paper refer only to step 3 and 4 as characteristic behavior, since all the other steps showed some transitory effect for the beginning and the end of the operation.

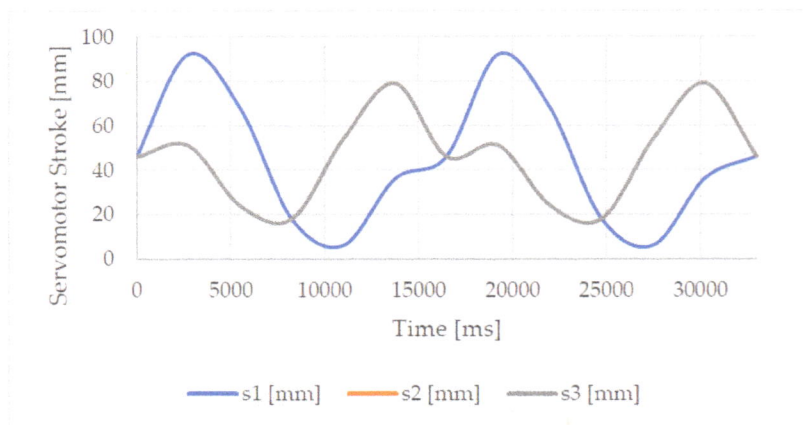

(a)

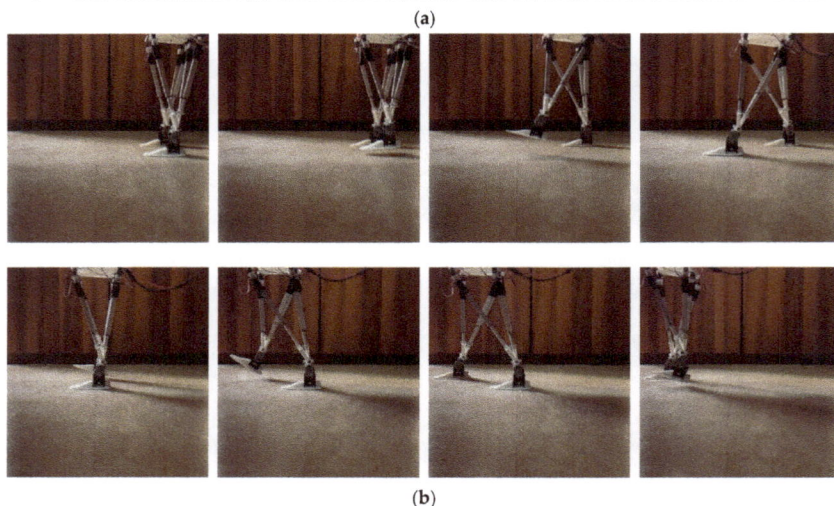

(b)

Figure 9. Walking test: (**a**) Input of the linear servomotors (s2 and s3 overlap); (**b**) Snapshots.

In Figure 10a, the results for the angular displacement of the foot during the step operation are shown. The measurements obtained with the IMU could be successfully compared with the expected angular displacement evaluated through a kinematic analysis and shown in Figure 10a in the form of a dotted line. A further characterization of the step operation could be obtained from the linear acceleration data acquired by the IMU, as reported in Figure 10b.

The acceleration was normalized to the standard acceleration due to gravity g (9.80665 m·s^{-2}), where the effect could be observed even on the value of the linear acceleration along the Z-axis, that was oscillating around 1g instead of 0. The graph in Figure 10b shows that there were no sudden changes in the motion and that the entire walking operation was fairly smooth, with a maximum acceleration of 0.2 g, thanks to the stiff operation of the parallel architecture of the leg design.

The power consumption of the locomotion system was obtained by multiplying the current measured by the current sensor and the voltage, fixed to 6.8 V by the power supply. The acquired data are shown in Figure 10c. The peak value of power consumption was lower than 11.000 W, whilst the power RMS (root mean square) was 2.536 W. Therefore, the system could be powered by a Li-Po battery, making the robot able to operate completely through Bluetooth or wireless technologies.

The estimated specific cost of transport [25] of the prototype, which was defined as cet = (energy used)/(weight)(distance traveled), was 2.5, which was comparable to the values for other biped robots as reported in Reference [25].

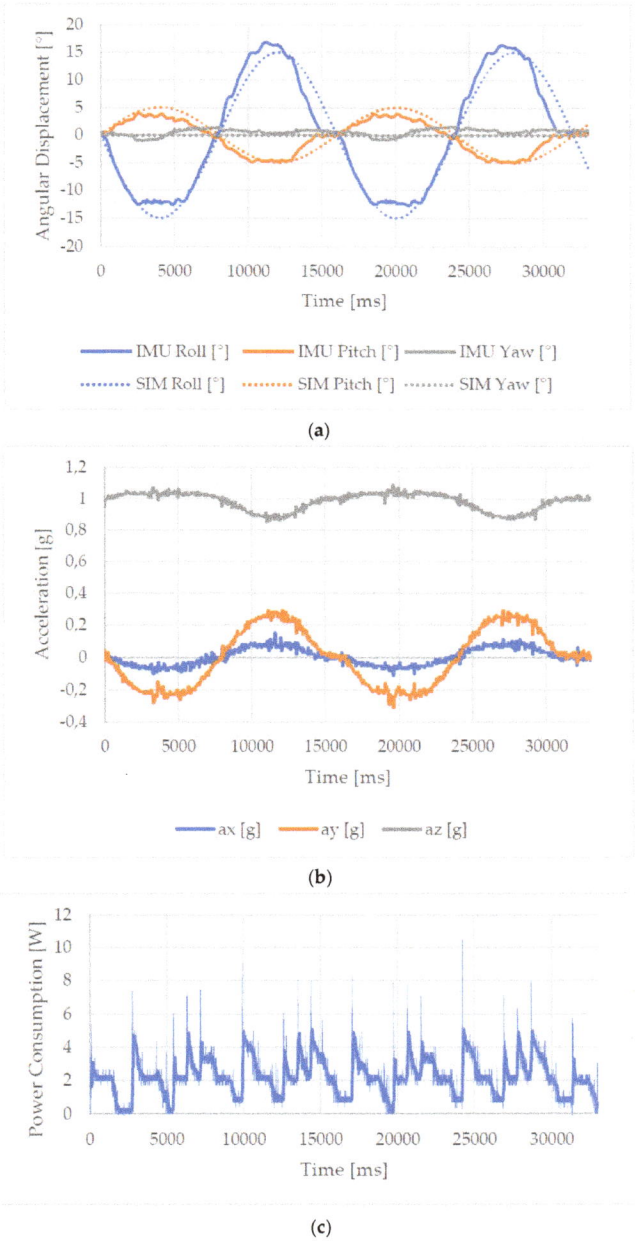

(a)

(b)

(c)

Figure 10. Results of a walking test: (**a**) angular displacement of the foot; (**b**) Accelerations measured by the IMU on the foot; (**c**) Acquired power consumption.

To prove the payload capability of the LARMbot 2 prototype, experimental load-lifting tests were carried out. The experiments are characterized by a lifting motion of LARMbot 2 while subject to its own weight, and an additional weight of 1.00 kg was applied to the back of the humanoid. Table 3 reports the main parameters for the two experimental modes. For both test modes, an IMU sensor in the neck of the prototypes measured angular displacement and linear acceleration. A current sensor was used to get the power, since the entire system was powered by a 12 V supply.

Table 3. Experimental test modes.

Test Mode	Payload	Appl. Point	Moving Parts	Sensors
1	0.00 kg	-	Legs, Torso	IMU, Current Sensor
2	1.00 kg	Back	Legs, Torso	IMU, Current Sensor

The first test mode was characterized by a lifting movement along the vertical axis equal to 80 mm. There was no payload applied to the structure apart from its own weight. The results for the first test mode are shown in Figures 11 and 12. The plots in Figure 11 showed that the torso tilted slightly forward during the lifting phase (pitch angle varying by approximately 4°), whilst there was a transient change of facing at the beginning and at the end of the lifting phase (yaw angle varying). The angular motion was extremely limited, with a maximum variation of 5°. The motion was performed in a smooth, continuous motion, as shown by the small variation in the linear acceleration in Figure 12. The peak of power consumption in this test was approximately 20.00 W.

The second test mode was characterized by the same lifting movement along the vertical axis. There was a payload of 1.00 kg applied to the back of the torso. The results for this test mode are shown in Figures 13 and 14. The plots in Figure 13 show that the torso tilts forward during the lifting phase (pitch angle varying of approximately 15°), while the variation of the other angles is less significant. The angular motion was extremely limited, with a maximum variation of 5°. The motion was performed in a smooth, continuous motion, as shown by the small variation in the linear acceleration in Figure 14. The peak of power consumption in this test was approximately 25.00 W.

When compared to the results of the previous test mode, the torso was tilted forward by 10° for the beginning to balance the weight on its back, and a bigger displacement on the pitch angle was needed to keep balance during motion. The acceleration plots of the second mode showed more disturbances than the first one, but the plots variation was contained within 2 m/s². Overall, the experimental tests validated the payload capability of the proposed structure.

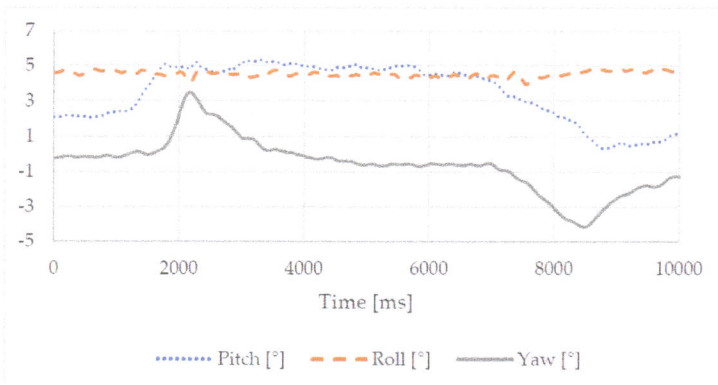

Figure 11. Acquired angular displacement for test mode 1 without payload.

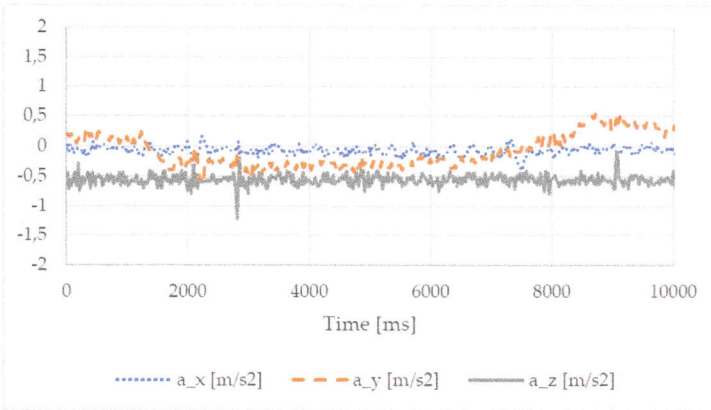

Figure 12. Acquired linear acceleration for test mode 1 without payload.

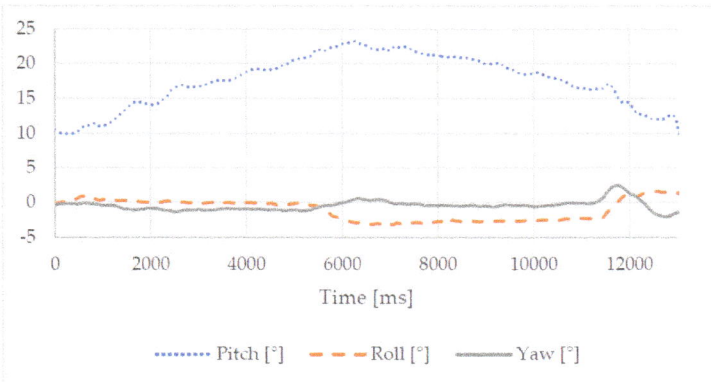

Figure 13. Acquired angular displacement for test mode 2 with payload.

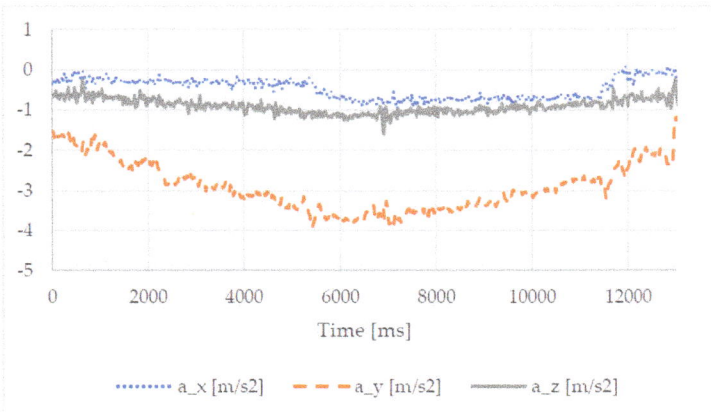

Figure 14. Acquired linear acceleration for test mode 2 with payload.

In conclusion, the experimental tests validated the expected performance of the LARMbot 2. The locomotion module could perform walking operations with a step size that was large when compared to other existing legs with parallel architecture, owing to the special joint mechanism first introduced in Reference [26]. Furthermore, the parallel architectures of the leg and torso allowed LARMbot 2 to lift a considerable weight. The power consumption of the whole humanoid was lower than 20 W for all the experimental tests.

6. Conclusions

A novel humanoid robot, the LARMbot 2, was introduced in this paper as a lightweight, low-cost solution based on parallel architectures for the torso and leg mechanisms. The kinematics of the whole humanoid was described, and a CAD model was presented for a constructive solution that was validated with a prototype. The parallel architectures on which the robot is based allowed it to lift a considerable payload, especially when compared to the weight of the entire system, as proven by the experimental data. Furthermore, the performance of the novel leg design was evaluated during a walking operation by both numerical computation and experimental tests. The results showed that the proposed leg mechanism could perform the walking task with good efficiency and low power consumption, with an improved performance compared to the previous LARMbot design. A full dynamic control that coordinates the legs, torso, and arms for dynamic walking and manipulation will be developed in future works.

7. Patents

Russo M., Cafolla D., Ceccarelli M., "Dispositivo per gamba tripode (Device for tripod leg)", IT patent application 102016000097258, 28/09/2016.

Russo M., Ceccarelli M., "Dispositivo di collegamento sferico tra tre corpi (Device for the spherical connection of three bodies)", IT patent application 102016000093695, 19/09/2016.

Author Contributions: Conceptualization, M.R. and M.C.; Data curation, M.R. and D.C.; Formal analysis, M.R.; Funding acquisition, M.C.; Investigation, M.R. and D.C.; Methodology, M.C.; Project administration, M.C.; Software, M.R. and D.C.; Supervision, M.C.; Validation, M.R. and D.C.; Writing—original draft, M.R.; Writing—review & editing, M.R. and M.C.

Funding: This research received no external funding.

Conflicts of Interest: The authors declare no conflict of interest.

References

1. Lim, H.O.; Takanishi, A. Biped walking robots created at Waseda University: WL and WABIAN family. *Philos. Trans. R. Soc. Lond. A* **2007**, *365*, 49–64. [CrossRef] [PubMed]
2. Kulk, J.; Welsh, J. A low power walk for the NAO robot. In Proceedings of the 2008 Australasian Conference on Robotics & Automation (ACRA-2008), Canberra, Australia, 3–5 December 2008; Kim, J., Mahony, R., Eds.; pp. 1–7.
3. Metta, G.; Natale, L.; Nori, F.; Sandini, G.; Vernon, D.; Fadiga, L.; Bernardino, A. The iCub humanoid robot: An open-systems platform for research in cognitive development. *Neural Netw.* **2010**, *23*, 1125–1134. [CrossRef] [PubMed]
4. Tsagarakis, N.G.; Caldwell, D.G.; Negrello, F.; Choi, W.; Baccelliere, L.; Loc, V.G.; Natale, L. WALK-MAN: A High-Performance Humanoid Platform for Realistic Environments. *J. Field Robot.* **2017**, *34*, 1225–1259. [CrossRef]
5. Lafaye, J.; Gouaillier, D.; Wieber, P.B. Linear model predictive control of the locomotion of Pepper, a humanoid robot with omnidirectional wheels. In Proceedings of the IEEE 2014 14th IEEE-RAS International Conference on Humanoid Robots, Madrid, Spain, 18–20 November 2014; pp. 336–341.
6. Ogura, Y.; Aikawa, H.; Shimomura, K.; Morishima, A.; Lim, H.O.; Takanishi, A. Development of a new humanoid robot WABIAN-2. In Proceedings of the IEEE International Conference on Robotics and Automation, Orlando, FL, USA, 15–19 May 2006; pp. 76–81.

7. Jung, H.W.; Seo, Y.H.; Ryoo, M.S.; Yang, H.S. Affective communication system with multimodality for a humanoid robot, AMI. In Proceedings of the 2004 4th IEEE/RAS International Conference on Humanoid Robots, Santa Monica, CA, USA, 10–12 November 2004; Volume 2, pp. 690–706.

8. Tellez, R.; Ferro, F.; Garcia, S.; Gomez, E.; Jorge, E.; Mora, D.; Pinyol, D.; Oliver, J.; Torres, O.; Velazquez, J.; et al. Reem-B: An autonomous lightweight human-size humanoid robot. In Proceedings of the IEEE 8th IEEE-RAS International Conference on Humanoid Robots, Daejeon, Korea, 1–3 December 2008; pp. 462–468.

9. Asfour, T.; Regenstein, K.; Azad, P.; Schroder, J.; Bierbaum, A.; Vahrenkamp, N.; Dillmann, R. ARMAR-III: An integrated humanoid platform for sensory-motor control. In Proceedings of the 6th IEEE-RAS International Conference on Humanoid Robots, Genova, Italy, 4–6 December 2006; pp. 169–175.

10. Asfour, T.; Schill, J.; Peters, H.; Klas, C.; Bücker, J.; Sander, C.; Bartenbach, V. Armar-4: A 63 DOF torque controlled humanoid robot. In Proceedings of the 13th IEEE-RAS International Conference on Humanoid Robots (Humanoids), Atlanta, GA, USA, 15–17 October 2013; pp. 390–396.

11. Cafolla, D.; Wang, M.F.; Carbone, G.; Ceccarelli, M. LARMbot: A new humanoid robot with parallel mechanisms. In *Romansy 21—Robot Design, Dynamics and Control: Proceedings of the 21st Cism-Iftomm Symposium, June 20–23, Udine, Italy*; Springer International Publishing: Cham, Switzerland, 2016; pp. 275–284.

12. Ceccarelli, M.; Cafolla, D.; Russo, M.; Carbone, G. LARMBot Humanoid Design towards a Prototype. *MOJ Appl. Bionics Biomech.* **2017**, *1*, 48–49. [CrossRef]

13. Cafolla, D.; Ceccarelli, M. Design and simulation of a cable-driven vertebra-based humanoid torso. *Int. J. Humanoid Robot.* **2016**, *13*, 1650015-1–1650015-27. [CrossRef]

14. Cafolla, D.; Ceccarelli, M. An Experimental Validation of a Novel Humanoid Torso. *Robot. Auton. Syst.* **2017**. [CrossRef]

15. Wang, M.; Ceccarelli, M. Design and simulation of walking operation of a cassino biped locomotor. In *New Trends in Mechanism and Machine Science*; Springer International Publishing: Cham, Switzerland, 2015; pp. 613–621.

16. Russo, M.; Ceccarelli, M.; Takeda, Y. Force transmission and constraint analysis of a 3-SPR parallel manipulator. *Proc. Inst. Mech. Eng. Part C J. Mech. Eng. Sci.* **2017**. [CrossRef]

17. Russo, M.; Herrero, S.; Altuzarra, O.; Ceccarelli, M. Kinematic Analysis and multi-objective optimization of a 3-UPR parallel mechanism for a robotic leg. *Mech. Mach. Theory* **2018**, *120*, 192–202. [CrossRef]

18. Cafolla, D.; Chen, I.M.; Ceccarelli, M. An experimental characterization of human torso motion. *Front. Mech. Eng.* **2015**, *10*, 311–325. [CrossRef]

19. Ceccarelli, M.; Carbone, G.; Cafolla, D.; Wang, M. How to Use 3D Printing for Feasibility Check of Mechanism Design. In *Advances in Robot Design and Intelligent Control*; Borangiu, T., Ed.; Advances in Intelligent Systems and Computing; Springer: Cham, Switzerland, 2016; Volume 371.

20. Arduino Nano Datasheet. Available online: https://store.arduino.cc/arduino-nano (accessed on 16 November 2018).

21. Adafruit PWM Servo Driver Datasheet. Available online: https://www.adafruit.com/product/815 (accessed on 16 November 2018).

22. Actuonix L12R Linear Servomotor Datasheet. Available online: https://www.actuonix.com/L12-R-Linear-Servo-For-Radio-Control-p/l12-r.htm (accessed on 16 November 2018).

23. MG 995 Servomotor Datasheet. Available online: https://www.towerpro.com.tw/product/mg995/ (accessed on 16 November 2018).

24. Adafruit IMU Datasheet. Available online: https://www.adafruit.com/product/3463 (accessed on 16 November 2018).

25. Collins, S.H.; Ruina, A. A bipedal walking robot with efficient and human-like gait. In Proceedings of the 2005 IEEE International Conference on Robotics and Automation, Barcelona, Spain, 18–22 April 2005.

26. Russo, M.; Ceccarelli, M. Kinematic design of a tripod parallel mechanism for robotic legs. In *Mechanisms, Transmissions and Applications*; Mechanism and Machine Science; Springer: Cham, Switzerland, 2018; Volume 52, pp. 121–130.

robotics

MDPI

Article

Non-Photorealistic Rendering Techniques for Artistic Robotic Painting

Lorenzo Scalera [1,2,*], Stefano Seriani [3], Alessandro Gasparetto [1] and Paolo Gallina [3]

[1] Polytechnic Department of Engineering and Architecture, University of Udine, 33100 Udine, Italy; alessandro.gasparetto@uniud.it
[2] Faculty of Science and Technology, Free University of Bozen-Bolzano, 39100 Bolzano, Italy
[3] Department of Engineering and Architecture, University of Trieste, 34127 Trieste, Italy; sseriani@units.it (S.S.); pgallina@units.it (P.G.)
* Correspondence: lorenzo.scalera@uniud.it

Received: 21 December 2018; Accepted: 7 February 2019; Published: 11 February 2019

Abstract: In this paper, we present non-photorealistic rendering techniques that are applied together with a painting robot to realize artworks with original styles. Our robotic painting system is called Busker Robot and it has been considered of interest in recent art fairs and international exhibitions. It consists of a six degree-of-freedom collaborative robot and a series of image processing and path planning algorithms. In particular, here, two different rendering techniques are presented and a description of the experimental set-up is carried out. Finally, the experimental results are discussed by analyzing the elements that can account for the aesthetic appreciation of the artworks.

Keywords: painting robot; collaborative robot; image processing; non-photorealistic rendering; artistic rendering

1. Introduction

Robotic painting is a challenging task that is motivated by an inner wish to discover novel forms of art and to experiment the technological advances to create something that can be aesthetically appreciated. Developing an automatic robotic painting system is hard, since the process comprises several different fields, including robotics, automation, image processing and art.

Busker Robot has been developed since 2016 at University of Trieste, Italy, in collaboration with University of Udine, Italy. The robotic system is composed of a six degree-of-freedom (DOF) robotic arm and a series of image processing and path planning algorithms that are capable of interpreting an input digital image into a real artwork. The aim of the work is, therefore, not to faithfully reproduce an image, as typical printers or plotters do, but to introduce an original contribution. The image processing is the result of the extensive implementation of non-photorealistic rendering (NPR) techniques with a trial and error procedure that is developed to the fulfillment of the artist. In each algorithm that we have implemented, several parameters can be controlled in order to modify the desired outcome. Moreover, these techniques are not fully deterministic, since random effects have been introduced, as explained in the paper, to obtain a different and non-repeatable result every time the algorithm is run.

The robotic machine that we have adopted for our purposes is a UR10 collaborative robot, produced by Universal Robots. Its collaborative features, such as force and speed limits as well as collision-detection systems, allow an artist to work side by side with the robotic arm. For example, the paint level, the type of brush and color can be changed or adjusted during the painting process, when needed.

Busker Robot, which name refers to street artists, has been previously presented in [1,2] and has been showcased for the first time in 2016 at the exhibition "Art, Science, Technology, Robotics" in Trieste, Italy. Then, it was shown at the SPS IPC Drives Italy 2017, at the "Algorithmic Arts and Robotics" exhibition during the international event Trieste Next 2017, at "Piccolo Teatro" in Milan, 2017, and, more recently, at the international festival "Robotics" in Trieste, 2018 (Figure 1) [3]. Furthermore, it took part to the 2018 International Robotic Art Competition (RobotArt), a context where 19 teams from all over the world competed and more than 100 artworks were created by painting robots; Busker Robot won an Honorable Mention [4].

Figure 1. Busker Robot at the international festival "Robotics" (Trieste, November 2018).

With respect to previously published works [1,2], in this paper: (a) we present *two novel non-photorealistic rendering techniques*, *cross-hatching* and Random Splines, which are adopted for the artistic rendering of light and shades on an image, and (b) we employ *non-diluted black Indian ink*, instead of the previously adopted watercolor and gouache techniques.

The change from watercolor to ink was prompted by the idea to experiment the filling of areas with patterns of thin lines, in order to obtain different effects from those generated by water and pigments in the watercolor and gouache techniques. With respect to watercolor, in which the brush precision is not a big issue since the effects generated in the watered paper are uncontrollable and random, by handling the ink, the accuracy is much more important. Moreover, the height of the brush, the interval between two consecutive color refills and the planarity of the paper are much more critical.

With the ink painting technique several layers of thin lines and curves are applied to the paper, each one overlapped to the previous one, in order to produce dark effects in the shaded areas and lightness in the areas that are not covered by strokes. Experimental results show the feasibility of the proposed approach and good performances of the system architecture.

In the next section we provide a brief background including related works, in Section 3 we describe the architecture of the painting system, in Section 4 we present the non-photorealistic rendering techniques, and in Section 5 we include a description of the hardware set-up. The results are reported in Section 6, whereas Section 7 highlights the conclusion of this work and possible future improvements.

2. Related Work

Since the middle of the twentieth century, when the first programmable machines were adopted and what is known as "industrial robotics" started [5], research in this field has been focused in developing manipulators capable of replacing human labor in hazardous environments, which increases workplace safety and enhances production levels. By contrast, in recent years, the original industrial use has evolved and several artists have brought robotic machines from the factories to public exhibition spaces to develop novel artistic forms.

One of the earliest painting machines can be identified in the draughtsman automaton by Pierre Jaquet Droz, built in the 1760s [6]. In the last century, Jean Tinguely (1925–1991) can be considered one of the first artists that created painting machines to produce artworks with mostly random patterns [7]. Harold Cohen (1928–2016) devised AARON, a plotter capable of reproducing computer-generated artworks, considered pioneer in algorithmic and generative art [8].

In more recent years, several artists, engineers and computer scientists developed robots and machines skilled at drawing and painting. One example is given by the humanoid drawing human portraits by S. Calinon et al. [9], whereas G. Jean-Pierre and Z. Said developed the Artist Robot, a 6-DOF industrial robot programmed for drawing portraits using visual feedback [10]. Other examples are given by the works of C. Aguilar et al., who developed a system composed of an articulated painting arm and a machine-learning algorithm aimed at determining a series of brush strokes to be painted on the canvas [11]. An interesting example of robotic installation capable of drawing sketches of people is given by the machine developed by P. Tresset and F. F. Leymarie [12,13]. The system, based on visual feedback to iteratively augment and improve the sketch, draws using the equivalent of an artist's stylistic signature. Furthermore, T. Lindemeier et al. presented e-David, a modified industrial robot that works with visual feedback and it is based on non-photorealistic rendering techniques [14,15]. The painting machine applies different layers of strokes on the canvas and produces beautiful artworks that mimic those produced by humans. Recently, the brush technique has been further improved by J. M. Gülzow et al. [16], who developed a method to analyze human brushstrokes and replicate them by means of the robotic painting system with impressive results.

More recently, examples of robots capable of producing artworks can be found in the work of X. Dong et al., who adopted a Scara robot to paint stylized portraits obtained with an algorithm based on a triangle coordinate system [17], of D. Berio et al., who adopted a collaborative Baxter robot to produce graffiti strokes [18], and by D. Song et al., who presented a 7-DOF robotic pen-drawing system, that is capable of creating pen art on arbitrary surfaces [19]. Furthermore, a robot artist performing cartoon style facial portrait painting using NPR techniques has been presented in [20].

Another application of robots in art, other than in pen drawing and brush painting, is that of spray painting, a task typically employed in industry [21–23]. Examples of robotic spray painting for artistic purposes can be found in the works of Viola Ago [24] and L. Scalera et al. [25], who adopted industrial robotic arms equipped with an airbrush, and in the work of A. S. Vempati et al., who developed PaintCopter, an autonomous UAV for spray painting on 3D surfaces [26].

Finally, other examples of artistically skilled robots are given by painting mobile platforms, such the ones by L. Moura, who developed Robotic Action Painter in 2006 [27], and by C.-L. Shih, who presented the trajectory planning and control of a mobile robot for drawing applications [28].

3. Busker Robot

Busker Robot is a system with amodular architecture, consisting of both hardware and software parameters (Figure 2). The hardware is composed of the painting machine, a 6-DOF robotic arm by Universal Robots, mounted in front of the painting support and equipped with a changeable brush. The software consists of the image processing and the path planning algorithms, that have been implemented in MATLAB®. The image processing algorithms, i.e., the non-photorealistic rendering techniques, are applied to a digital input image that is processed in order to extract the contours, the details and the backgrounds that have to be reproduced on paper. Several different algorithms have been developed for Busker Robot [1,2]: it is the artists responsibility to choose which technique, or combination of techniques, has to be applied case by case.

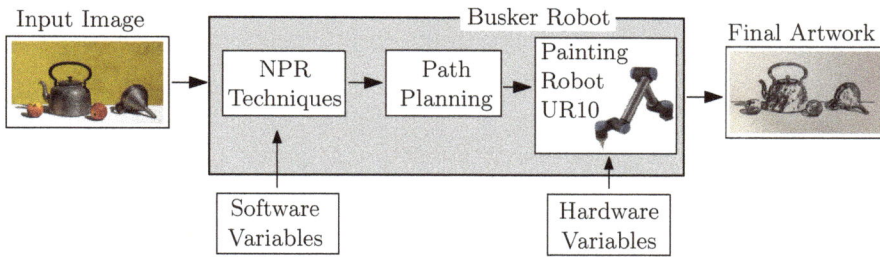

Figure 2. Overview on the system architecture.

The output of the artistic rendering is a structured list of points that identify the processed image. In particular, a MATLAB structure is adopted to handle the pixel coordinates of each single line or curve that composes the final image: each stroke is represented by an array that contains the coordinates of all the required way-points. These pixel coordinates are then scaled into the corresponding coordinates in millimeters on the painting canvas with a linear transformation. A trajectory planning is applied to these paths, in order to define the velocities profiles that the robot end-effector has to follow during the execution of the painting task. The trajectories are planned by adopting trapezoidal speed profiles for each motion: after the acceleration phase, the end-effector of the robot linearly moves with constant speed and circular blends through the way-points. After a stroke is completed, the robot tool z-coordinate is increased in order to lift the brush from the paper; it is then moved linearly at constant speed towards the first point of the next stroke but remaining at a higher z-coordinate; finally, it is lowered at paper level and the new stroke is started.

In the trajectory planning module, all the commands for the paint refill and the brush drying are as well defined. The output of this process is the list of commands for the robot, written in the specific UR Script Programming Language in a specific *script* file, which includes all the instructions needed for a specific layer of the artwork (Section 4). In particular, each command line corresponds to the motion of the robot end-effector through a specific way-point and it includes the Cartesian position coordinates in the operational space, the orientation of the tool in axis-angle representation, the velocity, the acceleration and the blend radius.

The execution of the painting task is then started and monitored from a remote computer connected to the robot controller via Ethernet. It is worth noting that the software is modular and, therefore, if any parameter regarding the painting set-up has to be changed, only the trajectory planning can be rerun, based on the same results of the artistic rendering module.

4. Non-Photorealistic Rendering Techniques

The two novel non-photorealistic rendering techniques implemented in this work are described in the following. These techniques can be adopted for the processing of the light and shades of an image in a manner that can enhance the aesthetic appreciation. Indeed, we would like to introduce a motor activity in the brush painting, so as to recall the gestures of a human artist in the robotic strokes. It has been demonstrated that the aesthetic experiences are enhanced in dynamic paintings that evoke a sense of movement and activate the brain area of visual motion during the observation [29,30].

Similarly to [1], the algorithms here presented are based on a thresholding process that is applied to a grey-scale version of the original image. For a given threshold I_T, i.e., the value of a specific grey intensity level (between 0 and 1), each pixel of the original image with intensity $I(P_i)$, is transformed into white if $I(P_i) \geq I_T$, or into black, otherwise. In this manner, the image can be decomposed into several layers, each one characterized by a different area to be painted, in a way similar to [31].

The algorithms presented in the following, *cross-hatching* and Random Splines, are applied to several layers of the same image. The paths resulting from the algorithms are then overlapped to obtain the final result. The NPR algorithms can be used for the processing of the areas with uniform grey intensity level of an image; furthermore, contours and details can also be added to the final result by adopting other techniques such as Canny Edge Detector, Hough Transform or Difference of Gaussians [1,2,32]. The Canny Edge Detector algorithm has been adopted in this work.

4.1. Cross-Hatching

The first NPR technique that we present in this paper is the *cross-hatching*. It consists of several layers of cross-hatch marks that allow to create nuanced differences in tone and light. The basic idea of hatching is to cover an area with parallel lines placed closely together: the areas where the hatching is placed will appear darker or in shadow, whereas the areas without hatching will appear illuminated. After the first set of lines, a second set of hatch marks can be placed on top, with a different orientation with respect to the first layer. Several sets of closely spaced lines can be drawn, corresponding to different thresholds. In this way, more saturated dark effects are obtained in the darkest side of the artwork. In the literature, examples of *cross-hatching* rendering can be found in [33–35].

In order to give the artwork greater vibrancy and to reproduce the gesture of an human painter, random effects have been implemented in our algorithm, so as to obtain small deviations (of the order of few millimiters) in the alignment and linearity of the single lines. These are generated by introducing a random perturbation in the longitudinal and transversal direction of a predefined number of points in the line, which becomes a poly-line. The points of the poly-line are then automatically blended to feed the robot with a smooth path. The maximum error for each of the two directions can be manually set before running the algorithm. The lines are finally filtered in order to remove the shortest segments which would otherwise be painted as dots, by the robot.

Two input images have been adopted for the evaluation of the algorithms: the Hydrodynamic Power Station and the Still Life with Tea Kettle, reported in Figure 3a,b. In this work, we used the Canny edge detector algorithm to define the contours for both the two images. Simulated results of the *cross-hatching* algorithms are reported in Figures 4 and 5. In these images, the thickness of the lines is not representative of the real thickness of the brushstrokes, which is given by the z-coordinate of the brush. The threshold and orientation employed for each layer of the images are reported in Tables 1 and 2. From the figures it can be seen that, by adopting higher thresholds, the total covered area increases, the shades are better rendered and the light contrast is more marked.

Table 3 reports the list of software parameters that can be manually tuned in the *cross-hatching* technique.

(a)   (b)

Figure 3. Original images. (**a**) Hydrodynamic Power Station (Trieste, Italy). (**b**) Still Life with Tea Kettle.

Figure 4. Hydrodynamic Power Station, simulations with *cross-hatching*, using the parameters reported on the left (**a**) and on the right (**b**) of Table 1.

Figure 5. Still Life with Tea Kettle, simulations with *cross-hatching*, using the parameters reported on the left (**a**) and on the right (**b**) of Table 2.

Table 1. Threshold and orientation for each layer of the Hydrodynamic Power Station, simulations with *cross-hatching*.

Layer	Figure 4a		Figure 4b	
	Threshold	Angle	Threshold	Angle
1	0.20	80°	0.30	70°
2	0.30	30°	0.40	40°
3	0.35	45°	0.45	50°

Table 2. Threshold and orientation for each layer of the Still Life with Tea Kettle, simulations with *cross-hatching*.

Layer	Figure 5a		Figure 5b	
	Threshold	Angle	Threshold	Angle
1	0.10	5°	0.25	85°
2	0.15	50°	0.30	80°
3	0.30	10°	0.35	50°
4	0.40	30°	0.40	10°
5	0.45	70°	0.45	5°

Table 3. Software variables for the non-photorealistic rendering techniques.

NPR Technique	Software Variables
Cross-hatching	- grey threshold - distance between lines - angle of orientation - maximum error in longitudinal direction - maximum error in transversal direction - minimum length of lines
Random Splines	- grey threshold - number of random points - size of the box - survival rate parameter - minimum length of lines

4.2. Random Splines

A different technique for the rendering of light and shades is Random Splines. Differently from other works in the field of NPR, where splines curves have been used to approximate the contours of the subjects to be painted, e.g., in [36], in this paper we adopt spline curves to fill areas characterized by uniform grey intensity level in a random manner. In this technique, a predefined number of points is randomly generated inside the area defined by a grey-scale threshold. For each point a survival rate is computed, which is proportional to the grey intensity of that point in the original image and to a constant defined by the user. A selection of the points is then applied in order to facilitate the survival of those positioned in the darkest side of the image.

A box centered in each point is generated and three random points are selected within the box boundaries. These points are interpolated with a spline curve, which is then sampled in order to save the sequence of way-points that the robot end-effector has to follow at constant linear speed during the painting task. The splines are sampled in the Cartesian space to obtain a good compromise between accuracy and number of points: a high number of points could, indeed, increase the computational effort of the robot controller and introduce vibrations in the tool if the distance between the points is smaller than the blend radius. A graphical example of the spline generation is shown in Figure 6.

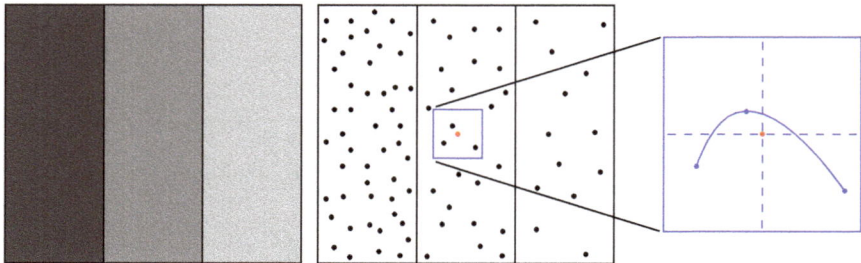

Figure 6. Example of spline generation in the Random Splines algorithm.

The size of the box allows to control the total length of the splines, whereas the survival rate constant allows to concentrate the strokes in the darkest part of the original grey-scale image. Since the points are placed inside a thresholded layer but the box is independent from the layer contours, the resulting splines can be drawn even in parts of the image that do not belong to that layer. This results in lines that exceed the natural border of an object in the image. The resulting splines can be finally filtered to remove those with the shortest path. The software parameters that can be manually defined are listed in Table 3. Two examples of Random Splines for the Still Life with Tea Kettle are shown in Figure 7, obtained with the parameters reported in Table 4.

(a)   (b)

Figure 7. Still Life with Tea Kettle, simulations with Random Splines, using the parameters reported on the left (**a**) and on the right (**b**) of Table 4.

Table 4. Threshold and box size (pixel) for each layer of the Still Life with Tea Kettle, simulations with Random Splines.

Layer	Figure 7a		Figure 7b	
	Threshold	Box Size	Threshold	Box Size
1	0.15	30	0.25	30
2	0.20	20	0.30	25
3	0.20	30	0.35	20
4	0.30	60	0.40	15
5	0.40	40	0.45	10
6	0.40	50	0.50	10

5. The Painting Machine

The painting machine that we adopted for Busker Robot is a 6-DOF UR10 robot. It is an industrial collaborative robot, usually adopted for pick-and-place, material handling, assembling or other manufacturing processes in environments where a human-robot interaction is needed. It is provided with speed and torque limits as well as collision-detection systems, that allow a human to work safely side by side with the machine. However, the speed of the robot has to be kept low, since it affects the quality of the painted strokes.

The UR10 can handle 10 kg of payload and can count on 1300 mm of working radius, which allows to easily paint on a 450 mm × 850 mm surface. Furthermore, a repeatability of ±0.1 mm on the positioning of the end-effector is ensured by the manufacturer [37]. The six axes allow to paint on non-horizontal surfaces and to perform complex motions such as dipping the brush in the paint cup or scrape it in order to remove excess paint. The robotic arm is fixed on an aluminum frame and its flange is equipped with a custom support for the application of the brushes by means of 3D printed plastic supports. An automatic brush change system has been developed for the adoption of brushes with different size, and used, in a previous work [1], to paint strokes with varied thickness.

The target surface for the painting task is identified with respect to the base coordinate system of the robot (Figure 8). Since a parallelism error between the base of the robot and the target surface could affect the application of color on the paper, a calibration of the painting board is carried out before starting a new artwork. Furthermore, if the target surface is not perfectly planar, a planarity error arises and an approximate interpolating plane has to be defined by more than 3 points. This calibration is performed by measuring a set of n points $P_i = \{x_i \ y_i \ z_i\}^T$, with $i = 1, \dots, n$, on the drawing surface, with respect to the robot reference frame $\{0\ 0\ 0\}^T$. After equipping the end-effector of the robot with a rigid and sharp tool with the same length of the brush, the tip of this is positioned so as to touch the surface and the corresponding pose of the robot is recorded. In order to calculate the approximating surface $Z = a_0 X + a_1 Y + a_2$, the following over-determined linear system has to be solved:

$$\begin{cases} a_0 x_1 + a_1 y_1 + a_2 = z_1 \\ \quad\quad\quad \ldots \\ a_0 x_n + a_1 y_n + a_2 = z_n \end{cases} \tag{1}$$

By rewriting the system in matrix form $AH = B$, with

$$A = \begin{bmatrix} x_1 & y_1 & 1 \\ \ldots & \ldots & 1 \\ x_n & y_n & 1 \end{bmatrix} \quad \text{and} \quad B = \begin{Bmatrix} z_1 \\ \ldots \\ z_n \end{Bmatrix} \tag{2}$$

The coefficient vector $H = \{a_0 \ a_1 \ a_2\}^T$ can be obtained as $H = A^+ B$, where $A^+ = (A^T A)^{-1} A^T$ is the pseudo-inverse matrix of A that minimizes the squared sum of errors. The z-coordinate of the robot tool is then automatically updated via software to paint on the approximating surface.

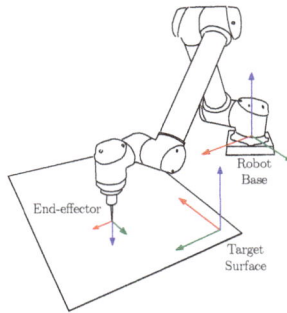

Figure 8. Robotic system reference frames.

Furthermore, the position of the paper, of the color cups and of the brush change repository is measured with respect to the robot reference frame, positioned in the central point of the robot base.

In this work, we adopted non-diluted black Indian ink provided to the robot in small plastic containers. After the dipping, the robot wipes off excessive paint from the brush on the edge of the ink cup.

A characterization of the brush strokes has been performed in order to estimate the influence of the z-coordinate of the brush (Figure 9), the traveled path and the robot speed on the stroke parameters: the black intensity and the thickness. This enables us to control the thickness of the strokes, to determine the traveled length after which the brush has to be dipped in the ink and to ensure a smooth deposition of the paint along the strokes.

To highlight the differences between watercolor and ink painting, some examples of stroke characterization are reported in Figures 10 and 11 for a watercolor and an ink stroke, respectively. It can be seen that in the ink case the black intensity is much more constant along the stroke, with respect to the watercolor one, where the dilution of pigments leads to transparency effects.

An experimental characterization has been performed and a set of linear strokes, each 250 mm long, has been painted by the robot while varying the z-coordinate from $z_1 = 0$ mm to $z_3 = -2$ mm with $\Delta z = 1$ mm, and the maximum robot speed, from $v_1 = 0.1$ m/s to $v_3 = 0.3$ m/s with $\Delta v = 0.1$ m/s. The maximum robot acceleration has been varied accordingly in order to maintain the same length of the constant velocity phase in the trapezoidal speed profile. Each stroke has been repeated three times and the results have been digitized and processed in order to obtain mean and standard deviation values. Figures 12 and 13 report these results, where the effects of the brush z-coordinate and robot speed on black intensity and stroke thickness can be seen, respectively.

From Figure 12, it can be seen that the black intensity is almost constant along the strokes and it is not affected by the z-coordinate or by the robot speed. The black intensity shows a limited increase in the last part of the stroke. This could be due to the trapezoidal motion profile, which is characterized by a deceleration phase that can affect the paint deposition process. In this manner, a slower speed can cause an increase in the paint deposition per linear distance. On the other hand, a high painting speed can cause undesirable effects in the starting and end points of the stroke, due to a rough landing and detachment of the brush on the target surface, as well as the dry brush effect [1].

Conversely, the stroke thickness shows a decreasing trend along the stroke (Figure 13). The z-coordinate of the brush affects the stroke thickness, since the more the bristles are pressed against the paper, the larger the strokes are painted. The robot speed seems not to be relevant in the tested range of values. The values of the mean thickness along the whole strokes have been computed as well and the results are reported as box-plots, in Figure 14.

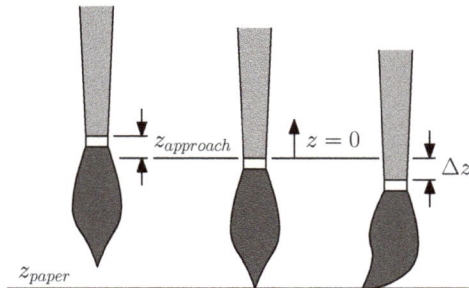

Figure 9. Positioning of the brush on the painting surface.

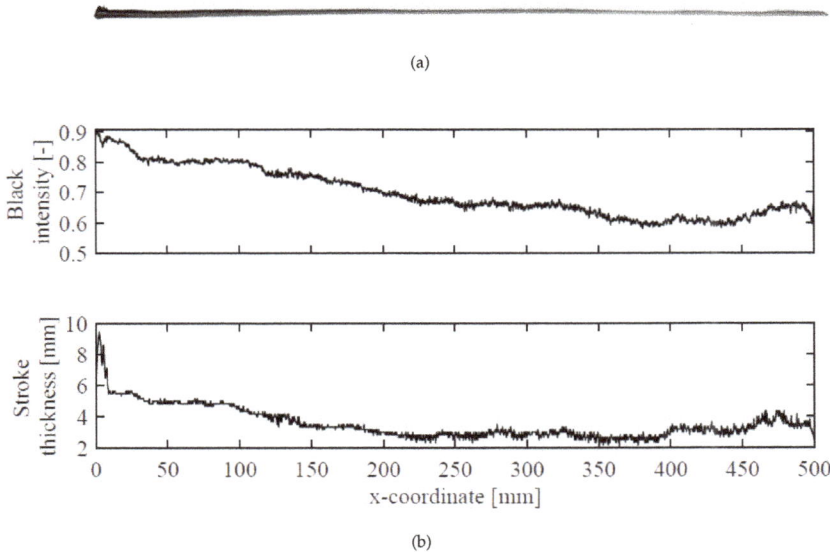

(a)

(b)

Figure 10. Example of *watercolor* stroke characterization: original photo (**a**), graphs of intensity and thickness (**b**).

(a)

(b)

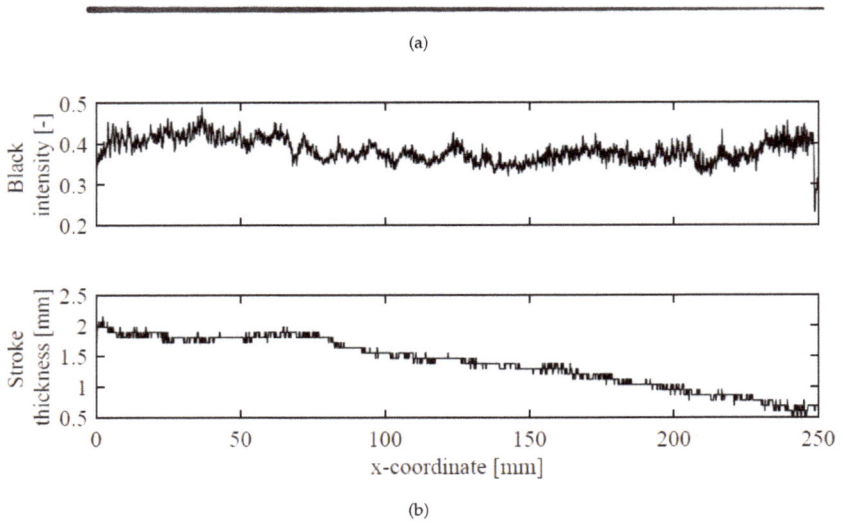

Figure 11. Example of *non-diluted ink* stroke characterization: original photo (**a**), graphs of intensity and thickness (**b**).

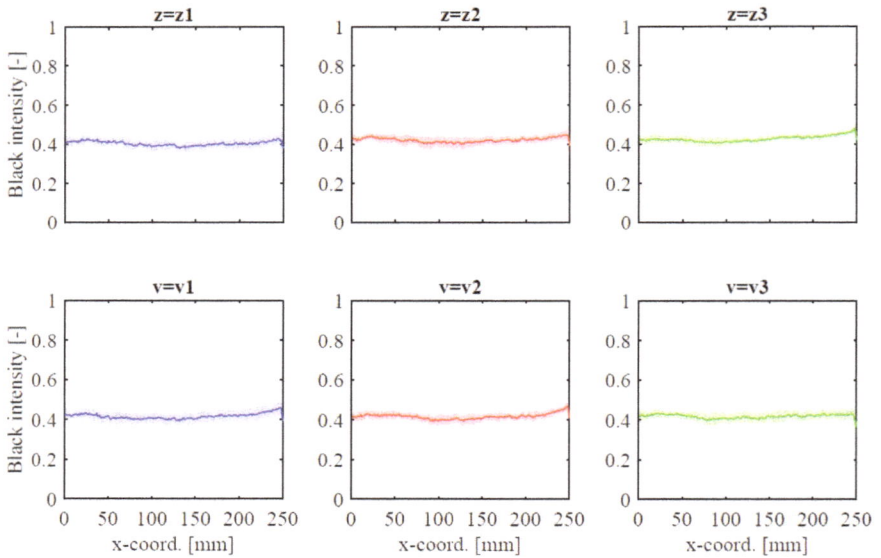

Figure 12. Effects of brush z-coordinate and robot speed on black intensity, mean and standard deviation values.

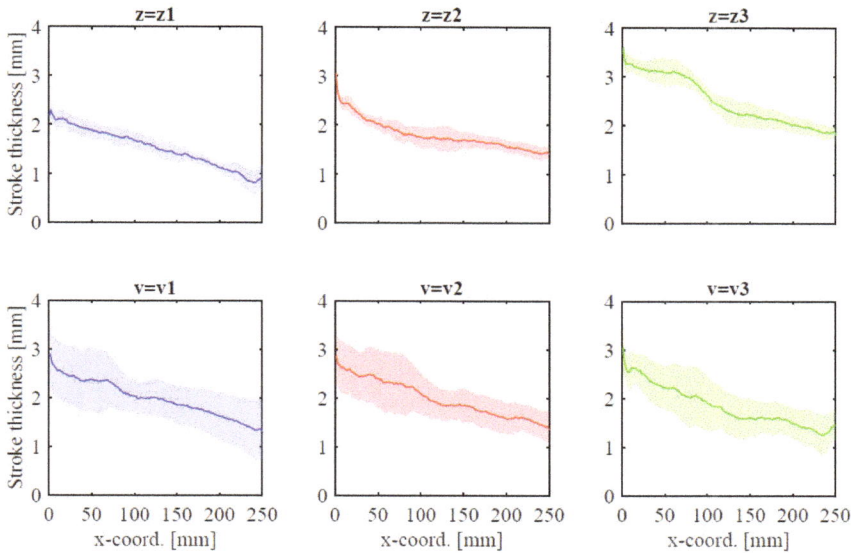

Figure 13. Effects of brush z-coordinate and robot speed on stroke thickness, mean and standard deviation values.

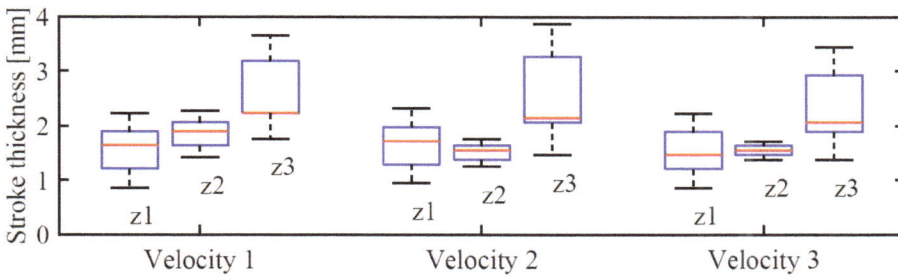

Figure 14. Box-plots of stroke thickness distribution: effects of robot speed and brush z-coordinate.

Busker Robot takes from one to three hours to complete one artwork, depending on the complexity of the subject, the number of layers and the speed of the robot. The total painting time can be optimized by finding the best compromise between quality of the strokes and speed of the robot, in order to minimize the brush travel time, while, at the same time, ensuring a smooth deposition of the paint along the strokes. A frame sequence of Busker Robot painting Hydrodynamic Power Station is reported in Figure 15.

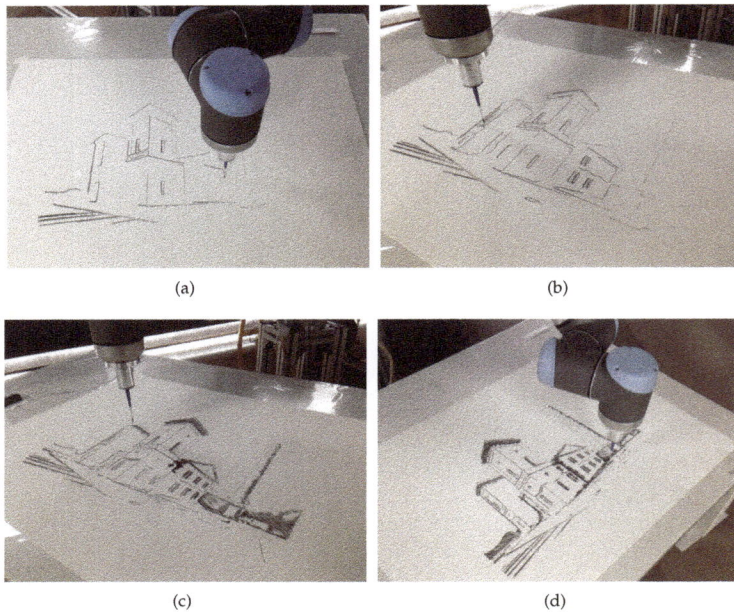

Figure 15. Busker Robot painting "Hydrodynamic Power Station". (**a**)–(**d**) show a frame sequence of the painting process in progress.

6. Results

Experimental results of the robotic painting system are reported in Figures 16 and 17, where the resulting artworks "Hydrodynamic Power Station" and "Still Life with Tea Kettle" are shown. The second subject has been realized with both the *cross-hatching* and the Random Splines techniques, to have a visual comparison between the results of the two NPR algorithms (Figure 17a,b).

The three artworks have been realized with the parameters reported in Tables 1, 2 and 4, respectively. While the *cross-hatching* reproduces the effects of the light and shades in the image with a clean and ordered style, the Random Splines approach creates a configuration of lines that are closer to a sketch drawn by a human. The lines exceed the contours and the result is not accurate and tidy as in the *cross-hatching* case.

Figure 16. Hydrodynamic Power Station, final artwork with *cross-hatching*.

(a)   (b)

Figure 17. Still Life with Tea Kettle, final artworks. (**a**) *Cross-hatching*. (**b**) Random Splines.

A precise calibration of the brush strokes is very challenging and, therefore, it can be seen that not all the strokes appear with the same thickness and color intensity. The interaction between brush bristles and paper, the ink deposition, the ripples of the paper and the amount of paint in the brush are not easy to be modeled and controlled, especially without feedback, and a lot of work can be done to improve the results. Nevertheless, the artworks produced by our automatic painting system are unique and can be considered an example of integration between robotics, image processing and art.

Even though a quantitative analysis of the artworks is not possible, since the qualitative comparison with the original image and the aesthetic appreciation is subjective, Busker Robot obtained a good feedback and has been considered of interest by the press and the public at several exhibitions.

7. Conclusions

In this paper we presented non-photorealistic rendering techniques for artistic robotic painting, that have been implemented for the processing of digital images into real artworks with Busker Robot, our robotic painting system. The system is capable of reproducing an image that has been previously processed through the application of artistic rendering algorithm. In particular, we presented two novel techniques, the *cross-hatching* and the Random Splines, that are used to process the light and shade of a painting. Experimental results shown the feasibility of the proposed approach and system architecture. In particular, two sample images have been tested with the new algorithms and the robotic system, with interesting results. Future works will include a further brush stroke characterization to further improve the quality and the repeatability of the lines. A visual feedback can as well be introduced to give the robotic system a consciousness of the current status of the painting, that can be used to correct possible errors or place the next stroke.

Author Contributions: Investigation, methodology and software, P.G. and L.S.; data analysis and processing, L.S. and S.S.; writing—original draft preparation, L.S.; writing—review and editing, L.S., S.S., P.G. and A.G.; supervision, P.G. and A.G.

Funding: This research was partially funded by Fondo Ricerca Ateneo, FRA 2015 (Internal Fund, University of Trieste), https://www.units.it/intra/ricerca/fra.

Acknowledgments: The authors would like to thank M. A. Grimaldi for his help in the development of the Random Splines algorithm.

Conflicts of Interest: The authors declare no conflict of interest.

References

1. Scalera, L.; Seriani, S.; Gasparetto, A.; Gallina, P. Watercolour Robotic Painting: A Novel Automatic System for Artistic Rendering. *J. Intell. Rob. Syst.* **2018**, doi:10.1007/s10846-018-0937-y. [CrossRef]

2.   Scalera, L.; Seriani, S.; Gasparetto, A.; Gallina, P. Busker robot: A robotic painting system for rendering images into watercolour artworks. *Mech. Mach. Sci.* **2019**, *66*, 1–8, doi:10.1007/978-3-030-00365-4_1. [CrossRef]

3.   Robotics, Festival of Art and Robotics, Trieste 2018. Available online: https://robotics.gruppo78.it/?lang= en (accessed on 18 December 2018).

4.   International Robotic Art Competition (RobotArt). Available online: http://robotart.org/archives/2018/ artworks/ (accessed on 18 December 2018).

5.   Gasparetto, A.; Scalera, L. From the Unimate to the Delta robot: the early decades of Industrial Robotics. In Proceedings of the 6th IFToMM International Symposium on History of Machines and Mechanisms, HMM 2018, Beijing, China, 26–28 September 2018.

6.   Pierre Jaquet Droz. Available online: https://history-computer.com/Dreamers/Jaquet-Droz.html (accessed on 10 January 2019).

7.   Jean Tinguely. Available online: https://en.wikipedia.org/wiki/Jean_Tinguely/ (accessed on 18 December 2018).

8.   Cohen, H. The further exploits of AARON, painter. *Stanford Hum. Rev.* **1995**, *4*, 141–158.

9.   Calinon, S.; Epiney, J.; Billard, A. A humanoid robot drawing human portraits. In Proceedings of the IEEE-RAS International Conference on Humanoid Robots, Tsukuba, Japan, 5–7 December 2005; pp. 161–166.

10.   Jean-Pierre, G.; Saïd, Z. The artist robot: A robot drawing like a human artist. In Proceedings of the 2012 IEEE International Conference on Industrial Technology, Athens, Greece, 19–21 March 2012; pp. 486–491.

11.   Aguilar, C.; Lipson, H. A robotic system for interpreting images into painted artwork. In Proceedings of the 11th Generative Art Conference, Milan, Italy, 16–18 December 2008; Volume 11.

12.   Tresset, P.A.; Leymarie, F. Sketches by Paul the robot. In Proceedings of the 8th Annual Symp. on Computational Aesthetics in Graphics, Visualization, and Imaging, Eurographics Association, Annecy, France, 4–6 June 2012; pp. 17–24.

13.   Tresset, P.; Leymarie, F.F. Portrait drawing by Paul the robot. *Comput. Graphics* **2013**, *37*, 348–363. [CrossRef]

14.   Lindemeier, T.; Metzner, J.; Pollak, L.; Deussen, O. Hardware-Based Non-Photorealistic Rendering Using a Painting Robot. *Comput. Graphics Forum* **2015**, *34*, 311–323. [CrossRef]

15.   Lindemeier, T.; Spicker, M.; Deussen, O. Artistic composition for painterly rendering. In Proceedings of the 21th International Symposium on Vision, Modeling and Visualization (VMV 2016), Bayreuth, Germany, 10–12 October 2016.

16.   Gülzow, J.; Grayver, L.; Deussen, O. Self-Improving Robotic Brushstroke Replication. *Arts* **2018**, *7*, 84. [CrossRef]

17.   Dong, X.; Li, W.; Xin, N.; Zhang, L.; Lu, Y. Stylized Portrait Generation and Intelligent Drawing of Portrait Rendering Robot. *DEStech Trans. Eng. Techno. Res.* **2018** [CrossRef]

18.   Berio, D.; Calinon, S.; Leymarie, F.F. Learning dynamic graffiti strokes with a compliant robot. In Proceedings of the 2016 IEEE/RSJ International Conference on Intelligent Robots and Systems (IROS), Daejeon, Korea, 9–14 October 2016; pp. 3981–3986.

19.   Song, D.; Lee, T.; Kim, Y.J.; Sohn, S.; Kim, Y.J. Artistic Pen Drawing on an Arbitrary Surface using an Impedance-controlled Robot. In Proceedings of the IEEE International Conference on Robotics and Automation (ICRA), Brisbane, Australia, 21–25 May 2018.

20.   Luo, R.C.; Liu, Y.J. Robot Artist Performs Cartoon Style Facial Portrait Painting. In Proceedings of the 2018 IEEE/RSJ International Conference on Intelligent Robots and Systems (IROS), Madrid, Spain, 1–5 October 2018; pp. 7683–7688.

21.   Trigatti, G.; Boscariol, P.; Scalera, L.; Pillan, D.; Gasparetto, A. A new path-constrained trajectory planning strategy for spray painting robots-rev.1. *Int. J. Adv. Manuf. Technol.* **2018**, *98*, 2287–2296, doi:10.1007/s00170-018-2382-2. [CrossRef]

22.   Gasparetto, A.; Vidoni, R.; Saccavini, E.; Pillan, D. Optimal path planning for painting robots. In Proceedings of the ASME 2010 10th Biennial Conference on Engineering Systems Design and Analysis, Istanbul, Turkey, 12–14 July 2010; pp. 601–608.

23.   Trigatti, G.; Boscariol, P.; Scalera, L.; Pillan, D.; Gasparetto, A. A look-ahead trajectory planning algorithm for spray painting robots with non-spherical wrists. *Mech. Mach. Sci.* **2019**, *66*, 235–242, doi:10.1007/978-3-030-00365-4_28. [CrossRef]

24. Viola Ago. Robotic airbrush painting. Available online: http://violaago.com/robotic-airbrush-painting/ (accessed on 18 December 2018).

25. Scalera, L.; Mazzon, E.; Gallina, P.; Gasparetto, A. Airbrush Robotic Painting System: Experimental Validation of a Colour Spray Model. In Proceedings of the International Conference on Robotics in Alpe-Adria Danube Region, Turin, Italy, 21–23 June 2017; pp. 549–556.

26. Vempati, A.S.; Kamel, M.; Stilinovic, N.; Zhang, Q.; Reusser, D.; Sa, I.; Nieto, J.; Siegwart, R.; Beardsley, P. PaintCopter: An Autonomous UAV for Spray Painting on Three-Dimensional Surfaces. *IEEE Rob. Autom. Lett.* **2018**, *3*, 2862–2869. [CrossRef]

27. Moura, L. A new kind of art: The robotic action painter. Available online: http://generativeart.com/on/cic/papersGA2007/16.pdf (accessed on 8 February 2019).

28. Shih, C.L.; Lin, L.C. Trajectory Planning and Tracking Control of a Differential-Drive Mobile Robot in a Picture Drawing Application. *Robotics* **2017**, *6*, 17. [CrossRef]

29. Ticini, L.F.; Rachman, L.; Pelletier, J.; Dubal, S. Enhancing aesthetic appreciation by priming canvases with actions that match the artist's painting style. *Front. Hum. Neurosci.* **2014**, *8*, 391. [PubMed]

30. Chatterjee, A.; Vartanian, O. Neuroscience of aesthetics. *Ann. N.Y. Acad. Sci.* **2016**, *1369*, 172–194. [CrossRef] [PubMed]

31. Seriani, S.; Cortellessa, A.; Belfio, S.; Sortino, M.; Totis, G.; Gallina, P. Automatic path-planning algorithm for realistic decorative robotic painting. *Autom. Constr.* **2015**, *56*, 67–75. [CrossRef]

32. Canny, J. A computational approach to edge detection. *IEEE Trans. Pattern Anal. Mach. Intell.* **1986**, *6*, 679–698. [CrossRef]

33. Winkenbach, G.; Salesin, D.H. Rendering parametric surfaces in pen and ink. In Proceedings of the 23rd annual conference on Computer graphics and interactive techniques, New Orleans, LA, USA, 4–9 August 1996; pp. 469–476.

34. Praun, E.; Hoppe, H.; Webb, M.; Finkelstein, A. Real-time hatching. In Proceedings of the 28th annual conference on Computer graphics and interactive techniques, Los Angeles, CA, USA, 12–17 August 2001; p. 581.

35. Zander, J.; Isenberg, T.; Schlechtweg, S.; Strothotte, T. High quality hatching. *Comput. Graphics Forum* **2004**, *23*, 421–430. [CrossRef]

36. Lewis, J.P.; Fong, N.; XueXiang, X.; Soon, S.H.; Feng, T. More optimal strokes for NPR sketching. In Proceedings of the 3rd International Conference on Computer Graphics and Interactive Techniques in Australasia and South East Asia, Dunedin, New Zealand, 29 November–2 December 2005; pp. 47–50.

37. Universal Robot UR10 Technical Details. Available online: https://www.universal-robots.com/media/1801323/eng_199901_ur10_tech_spec_web_a4.pdf (accessed on 10 January 2019).

robotics

MDPI

Article

A Comparison of Robot Wrist Implementations for the iCub Humanoid [†]

Divya Shah [1,2,*,‡], Yuanqing Wu [3], Alessandro Scalzo [1], Giorgio Metta [1] and Alberto Parmiggiani [1]

[1]   iCub Facility, Fondazione Istituto Italiano di Tecnologia, 16163 Genova GE, Italy; alessandro.sclazo@iit.it (A.S.); giorgio.metta@iit.it (G.M.); alberto.parmiggiani@iit.it (A.P.)
[2]   Dipartimento di Informatica, Bioingegneria, Robotica ed Ingegneria dei Sistemi [DIBRIS], Universitá degli Studi di Genova, 16145 Genova GE, Italy
[3]   Department of Industrial Engineering, University of Bologna, 40136 Bologna BO, Italy; yuanqing.wu@unibo.it
*   Correspondence: divya.shah@iit.it; Tel.: +39-339-414-3772
[†]   This paper is an extended version of our paper published in the Proceedings of the 2018 4th IFToMM Symposium on Mechanism Design for Robotics, Udine, Italy, 11–13 September 2018; titled *Comparison of Workspace Analysis for Different Spherical Parallel Mechanisms*; pp. 193–201.
[‡]   Current address: Via San Quirico 19, 16163 Genova, Italy.

Received: 21 January 2019; Accepted: 13 February 2019; Published: 17 February 2019

Abstract: This article provides a detailed comparative analysis of five orientational, two degrees of freedom (DOF) mechanisms whose envisioned application is the wrist of the iCub humanoid robot. Firstly, the current iCub mk.2 wrist implementation is presented, and the desired design objectives are proposed. Prominent architectures from literature such as the spherical five-bar linkage and spherical six-bar linkage, the OmniWrist-III and the Quaternion joint mechanisms are modeled and analyzed for the said application. Finally, a detailed comparison of their workspace features is presented. The Quaternion joint mechanism emerges as a promising candidate from this study.

Keywords: robot wrists; spherical parallel mechanism; orientational mechanisms; computer-aided design; workspace analysis; iCub

1. Introduction

Closed-chain mechanisms, particularly parallel mechanisms, are reputed to exhibit favorable characteristics with respect to their serial counterparts, mainly due to the possibility of distributing the load on the output member to several kinematic chains assembled in parallel and reducing moving inertia by locating the motors on or close to the fixed frame. Their potential advantages include: a larger payload to robot weight ratio, greater stiffness, better accuracy, and higher dynamic performance. Common drawbacks are a lower dexterity, a smaller workspace, complex kinematic geometry, and existence of singular configurations.

While the synthesis and optimization of translational parallel manipulators is a well understood problem that has been addressed in several works [1–3], the conceptual design of orientational parallel mechanisms with a large rotation range remains a challenging task. In this article, the practical implementation of this class of mechanisms is considered for the wrist design of humanoid robots. The reference application here is the iCub, a 53DOF open-source humanoid robot developed to support research in embodied cognition [4].

There has been significant research towards the design of robotic wrists over the years and the literature is rather large [5,6]. Early studies presented the use of a redundant spherical wrist with four converging revolute (R) joint serial chain; kinematically equivalent to a spherical joint [7,8].

A conceptual design to achieve unbounded joint motions by replacing the intermediate joint of a Euler-angle wrist with a four-bar linkage was proposed in [9], but its practical implementations showed considerable restrictions on the workspace. The "standard" two-axis gimbal system tends to be one of the predominant choices for its wide range of decoupled yaw/pitch motions, fully isotropic workspace and a straightforward kinematics [10,11]. Since traditional layouts are not suitable for the iCub because of volume limitations, the implementation of an orientational parallel mechanism was brought into consideration for the robot's wrist.

The humanoid robotics literature is rich of examples of 2DOF mechanisms with parallel kinematics, based on linear actuators; among these we can cite the wrist of the robot AILA [12], the ankle of WABIAN-2RIII [13] and the wrist of Roboray [14]. Preliminary implementations showed that this class of mechanisms is not viable for the iCub wrist, mainly for the following three reasons: (i) the large volume occupied by the linear ball-screw stages, (ii) limited rotation range due to the mechanism's self-collisions and iii) the presence of kinematic singularities in the workspace.

The focus was then shifted to a class of fully decoupled 2DOF PKMs that provide hemispherical workspace. Spherical linkage mechanisms such as the spherical five-bars [15] and spherical six-bar mechanisms [16,17], have all the revolute joint axes intersecting at a common point, thus promising more uniform kinematic behavior.

Another one of the most prominent works, was the OmniWrist-III [18] mechanism by *Ross-Hime Designs, Inc.*, which falls under the class of N-UU mechanism. Each limb of the mechanism comprises a pair of universal joints, which is mirror symmetric about a common plane [19,20]. In comparison to a single universal joint which is a Euler-angle mechanism, a N-UU mechanism works under the same principle of a homokinetic joint or coupling [19,21], and can be effectively analyzed using Lie group methods [22,23]. It is shown to have large workspace, hemispherical rotation capability, and slender form factor for the overall system.

Recently, Kim et al. reported on their implementation of the Quaternion joint [24], a design similar to the one patented by Lande and David in 1978 [25]. This has a 2 DOF joint emulating spherical pure rolling motion and is surrounded by two pairs of actuating wires, the motions of which directly correspond to the Quaternion values of the joint.

This article is further structured as follows: Section 2 discusses various strategies of actuator relocation to reduce the motor power requirements. The iCub wrist mk.2 design is presented in Section 3 and the desired design objectives are proposed in Section 4. Section 5 describes the computer-aided design (CAD) modeling and simulation of the selected mechanisms form the ones mentioned previously and Section 6 illustrates the various couplings between the workspace features and the joint angles obtained from the simulations. The obtained analyses are further discussed in Section 7 and concluded in Section 8.

2. Actuator Relocation

Most serial robotic manipulators comprise six or more DOF to provide complete control of the position in space and orientation of the end-effector. In most robots a functional distinction between the function of the DOF can be observed. The first three or four, most proximal DOF are generally employed to move the robot end-effector in space, while the distal DOF are used to orient the end-effector. The proximal and distal robot links and DOF are thus often loosely referred to as respectively the "arm" and "wrist" (Figure 1a). Given their position, it is of the utmost importance for robotic wrists to be light-weight because distal masses increase the power requirements of proximal DOF. A possibility to overcome this shortcoming is to relocate the wrist actuators to more proximal locations. In electrically actuated robots (The current article focuses on electrically actuated robots since the vast majority of autonomous robots that have demonstrated practical capabilities are electrically actuated; similar considerations nevertheless hold for other actuation technologies like hydraulics), conceptually there are three main ways to achieve this goal. These approaches are illustrated in Figure 1b–d where mechanisms are represented as planar for clarity.

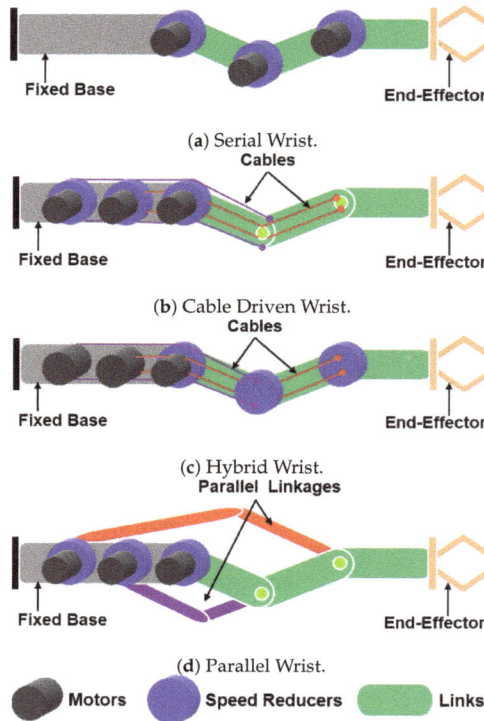

Figure 1. Conceptual actuator relocation configurations for wrist.

The first one, represented in Figure 1b is to place the motors fixed to the frame of previous links, and to convey the motive power to the wrist joints through a transmission system. Because of the complex rotations of wrist systems cable transmissions are generally adopted. This solution, for example, is employed in the wrist of the iCub robot (see Section 3). A drawback of this solution is that the use of cables introduces elasticities which, in turn, complicates the accurate control of the system.

The second one, represented in Figure 1c is to separate motors and speed-reducers, to keep the speed-reducers on the driven DOF, but to place the motors on proximal links and to connect them with fast, low force transmissions. Many authors have followed this approach; one of the first implementations, dating back to 1989, is the elbow mechanism of the Whole-Arm Manipulator (WAM) proposed by *Barrett Technologies* and later developed by Townsend and Salisbury [26]. More recent examples can be found in the work by Seok et al. [27] and of Kim on the LIMS robot arm [28].

The above two approaches will, however, inevitably increase the mechanical complexity of robots. Moreover, additional components are generally needed, which add to the total mass. Therefore, designers often face a delicate trade-off in striking a balance between adding masses (and complexity) for the transmissions, to reducing distal masses hence improving the functionality of a system. Also, these approaches are technically simpler in the case of planar motions. Unfortunately, most robots require non-planar joint arrangements.

A third alternative, represented in Figure 1d, is to achieve mass relocation by combining adjacent joints into multi-DOF (degree of freedom) parallel kinematics mechanisms. Examples of this approach can be found in [16,18,24,29]. A typical characteristic of parallel mechanisms is that their kinematic behavior tends to be more complex, often "non-uniform" (see [30,31]) with respect to their serial counterparts. This complicates both the design and control of this type of mechanisms.

This work compares the kinematic behavior of four such parallel mechanisms with large and regular workspaces, with that of the iCub wrist mk.2 (belonging to the category of Figure 1b) and the serial 2DOF gimbal mechanism (Figure 1a category) that are considered as a reference. The analyses focused on mechanisms with rotational actuators as inputs, although the authors envision extending this work to cover mechanisms with linear actuators as inputs (e.g., see [12,32]).

3. iCub Wrist mk.2

The hand-forearm assembly of the iCub humanoid robot [4], has 12 independent DOF, weighs 0.95 kg and has a volume of approximately 290 mm × 70 mm × 40 mm. These characteristics allow considerable dexterity, which comes, however, at the price of a limited robustness and great mechanical complexity. Significant amounts of efforts were devoted in recent years to improving the dependability of this sub-assembly, starting from the hand sub-system (e.g., see [33]). The current article instead, shifts the focus to the wrist.

The iCub wrist mk.2 is a 2DOF cable driven mechanism, as shown in Figure 2. The wrist is actuated by two Faulhaber 1331T012SR brushed DC motors, coupled to 159:1 planetary gear-heads that drive the pitch and yaw rotations of the hand. The motive power is transmitted by means of a cable-drive system (as represented in Figure 2b,c). The motor pulley and the driven pulley have slightly different diameters resulting in a 1.38 transmission ratio.

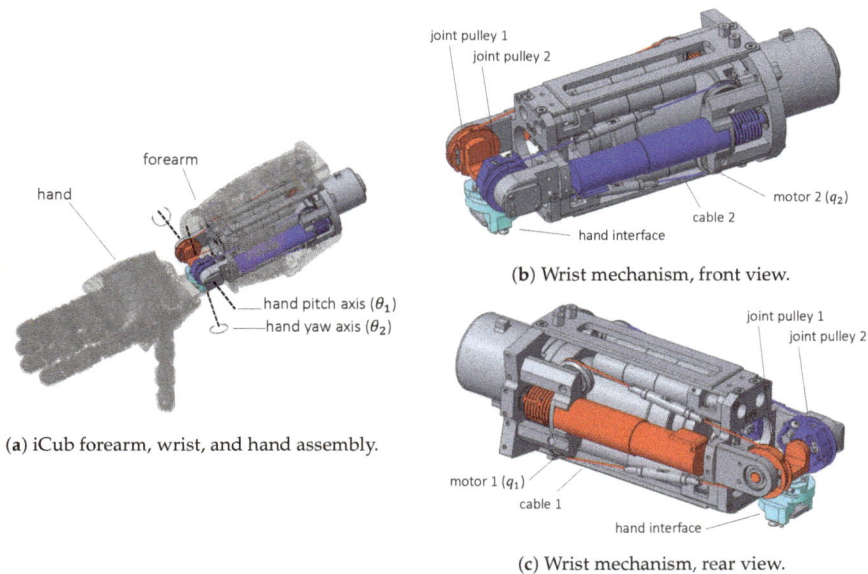

(b) Wrist mechanism, front view.

(a) iCub forearm, wrist, and hand assembly.

(c) Wrist mechanism, rear view.

Figure 2. Computer-Aided Design (CAD) representation of the iCub wrist.

The motion of motor 2 is transmitted to the hand yaw joint with a secondary cable system which conveys motion to the hand interface (the cyan part in Figure 2). This coupling is represented in the diagram of Figure 3, and gives rise to the following relation between the motor positions $q_m = [q_1, q_2]^T$ and hand orientations $\theta_j = [\theta_1, \theta_2]^T$:

$$\begin{bmatrix} \theta_1 \\ \theta_2 \end{bmatrix} = \begin{bmatrix} 1.38 & 0 \\ 1.38 & 1.38 \end{bmatrix} \begin{bmatrix} q_1 \\ q_2 \end{bmatrix} \tag{1}$$

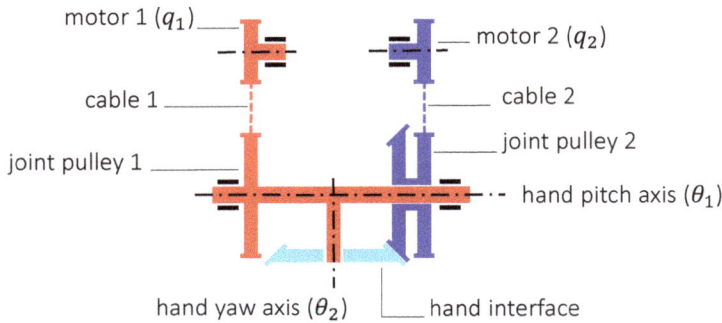

Figure 3. iCub wrist kinematic layout.

The characteristics of the wrist mechanism are summarized in Table 1.

Table 1. iCub Wrist Mechanism Parameters.

Parameter	Value
Max. continuous actuator torque	0.29 [Nm]
Max. actuator no load velocity	62.2 [rpm]
Max. continuous joint torque (pulleys)	0.40 [Nm]
Max. joint no load velocity	86.2 [rpm]
Joint 1 ROM	$[-56°, +56°]$
Joint 2 ROM	$[-38°, +38°]$

4. Design Objectives

As mentioned in the previous sections a promising way to overcome the shortcomings of the current iCub mk.2 wrist implementation is to consider a new wrist design, to improve robustness, and dependability. Furthermore, the elasticity of the current cable-drive system is detrimental for the control of the system and should be eliminated of possible.

Alternative wrist implementations should improve upon the baseline defined by Table 1, while fulfilling the following design criteria:

- **2 DOF:** The mechanism shall possess two DOF (e.g., a pitch and yaw motion along the two Cartesian axes). The 3rd DOF for the wrist (roll) is obtained at the level of the elbow of the robot and its mechanics are housed within the forearm. The wrist roll is not considered in the current analysis.
- **Large Range of Motion (ROM):** The mechanism shall possess a full hemispherical workspace, that is, a range of motion, possibly in the order of $\pm90°$ for each of the DOF.
- **Singularity-Free:** The workspace of the mechanism should be free of singularities, thus allowing a highly uniform or isotropic behavior of the mechanism throughout the workspace.
- **Full Decoupling:** An important feature for the mechanism is to have a decoupled motion, i.e., that the motion of one actuator results in the motion of one DOF independently from of the other, thus simplifying controller synthesis.
- **High Isotropy:** The Jacobian matrix for the mechanism should be constant and equal to identity throughout the workspace to allow easier control implementations.
- **Compact Design:** The maximum volume occupied by the wrist sub-assembly should be compact enough and compatible with the current hand-forearm assembly of the iCub humanoid, thus allowing easy integration of the new wrist into the system. The available volume can be approximated as a truncated cone with top and base diameters of 70 mm and 50 mm respectively and a height of 150 mm.

- **High Payload-to-Weight Ratio:** The moving mass of the mechanism must be minimized (or relocated in the proximal part) to allow manipulation of heavier payload with limited motor power/torque.

The current study presents the analyses of the 2DOF gimbal mechanism and the iCub mk.2 wrist mechanisms that are presented, for reference, as benchmarks. Besides these mechanisms, four alternative parallel mechanisms are considered:

- a spherical parallel mechanism with five curved links (bars) adapted from the one presented in [15];
- a spherical six bar mechanism as proposed in [16]
- an implementation of a N-UU parallel mechanisms similar to the OmniWrist-III mechanism [18] developed by *Ross-Hime Designs, Inc.*;
- a Quaternion joint, similar to the N-UU class, as proposed by Kim in [24] for the LIMS2-AMBIDEX robot.

5. CAD Modeling and Simulation

One of the drawbacks with the PKMs is that their kinematic relations are intricate and obtaining closed-form analytical solutions is rather complex. Thus, a CAD approach was followed to expedite the modeling and analysis process of the mechanisms. For each of the candidate mechanism, a CAD model of the kinematic architecture was developed using PTC Creo Parametric 4.0 and its Mechanism multi-body module. The workspace of the mechanisms was spanned by considering a mesh grid of all actuator input combinations within their admissible range. During the simulation, the in-built solvers from Creo compute the forward/inverse kinematics of the mechanism based on the modeled CAD structure for each of the input grid points. The simulation fails in case of any singularities and the does not produce a result for respective grid point. The resulting platform coordinates and orientation angles for the corresponding grid points were recorded from the simulation and later extracted for the analyses.

Indeed, the CAD-based method proved to be extremely convenient for rapidly assessing the workspace properties of the mechanism. Also, the CAD-based analysis can be very helpful in visualizing and detecting possible collisions and thus accelerating the overall design process.

To have a homogenized form factor for the mechanisms, based on the design objectives, all mechanism dimensions have been scaled to obtain a unit distance from the origin to the end-effector. This allows the workspace features of the mechanisms to be represented in an adimensional fashion. Consequently, in the following subsections, lengths will not be associated with their natural measurement units.

5.1. Gimbal

The 2DOF gimbal mechanism is a standard serial chain mechanism with two revolute joint axes successively placed along the two Cartesian axes, as shown in Figure 4a. The axes of the two actuators q_1 and q_2 lie along the Z-axis and Y-axis respectively, of the base frame attached to the fixed point O and result in the yaw and pitch motions.

5.2. iCub mk.2 Wrist

The structure of the iCub mk.2 wrist (Figure 4b) was described in detail in Section 3. The pitch and yaw motions are along the Y-axis and Z-axis, respectively.

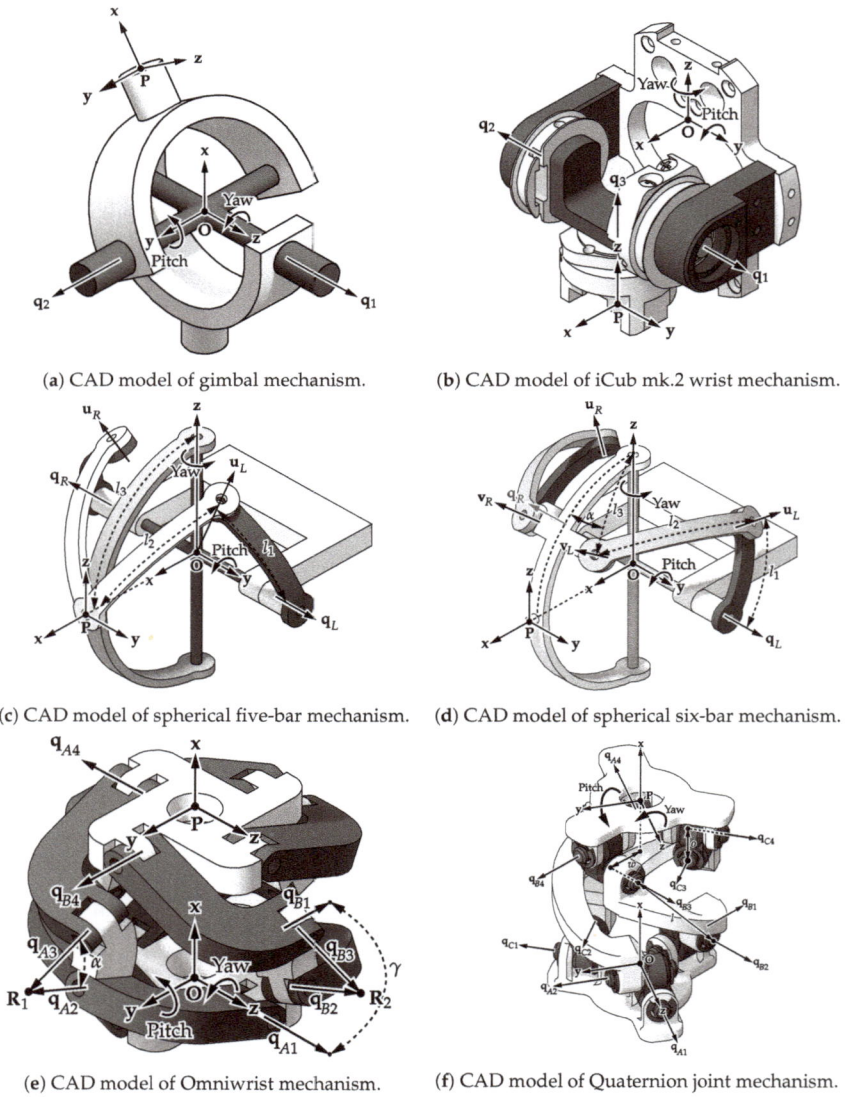

(**a**) CAD model of gimbal mechanism.

(**b**) CAD model of iCub mk.2 wrist mechanism.

(**c**) CAD model of spherical five-bar mechanism.

(**d**) CAD model of spherical six-bar mechanism.

(**e**) CAD model of Omniwrist mechanism.

(**f**) CAD model of Quaternion joint mechanism.

Figure 4. Computer-Aided Design (CAD) models for the mechanisms in consideration.

5.3. Spherical Five-Bar Linkage

The spherical five-bar mechanism has a kinematic chain of five revolute joints connected with curved linkages. Figure 4c shows a CAD model for this mechanism. All the axes of the mechanism intersect at the common central point O and the mechanism is symmetric with regard to the XZ-plane. The two actuation joints are attached diametrically opposite to the fixed base, and are indicated as q_L and q_R in Figure 4c. The joints u_L and u_R are passive. The end-effector point P undergoes pitch and yaw motions about the Y-axis and Z-axis respectively of the base frame attached to the fixed point O. It should be noted here that the mechanism has an additional constraint limb with a passive gimbal to restrict the parasitic roll motion of the end-effector. However, for simplicity, within this work the mechanism is referred to as a 'spherical five-bar mechanism'.

The parameter l_1 represents the angle between the Y-axis and the line along the joint u_L, parameter l_2 represents the angle between the line along u_L and the end-effector point P and the parameter l_3 represents the angle between the Z-axis and the end-effector point P. Starting from the geometric parameters proposed by the respective authors [15], the values were tweaked to suit the current application and were set to be $l_1 = 60°$, $l_2 = 74°$ and $l_3 = 90°$.

5.4. Spherical Six-Bar Linkage

The spherical six-bar mechanism is a spherical mechanism composed of six revolute joints and interconnected with curved links [16,17]; its CAD model is represented in Figure 4d. Similar to the spherical five-bar, the 'spherical six-bar mechanism', also has the additional constraint limb with a passive gimbal and it follows the similar nomenclature for the joint axes and frames. All the axes of the mechanism intersect at the common central point O and the mechanism is symmetric with regard to the XZ-plane. The actuated joints are q_L and q_R and the passive joints here are u_L, v_L, u_R and v_R. The pitch and yaw motions are along the Y-axis and Z-axis, respectively.

The parameter l_1 represents the angle between the Y-axis and the line along the joint u_L, parameter l_2 represents the angle between the lines joining u_L and v_L, l_3 corresponds to the angle between the line joining v_L and the Z-axis and the additional parameter α here, corresponds to the angle between l_3 and the XZ-plane. The parameter values were set to an optimal solution computed by differential evolution as proposed in [17]; $l_1 = 33.7°$, $l_2 = 83°$, $l_3 = 32.7°$, and $\alpha = 10.7°$.

5.5. OmniWrist-III (4-UU)

The OmniWrist-III mechanism is an N-UU type PKM with a moving platform connected to a fixed base through three or four identical limbs, each comprising of a serial chain of four non-coplanar revolute joints (RRRR) or equivalently two universal joints (UU). Figure 4e represents the CAD model for the 4-UU mechanism with joint angles q_{Lj}, $\forall L = A, B, C, D$ limbs and $\forall j = 1, ..., 4$ joints. The axes of rotation of the first two joints of each limb intersect at point O, the center of the fixed base. The axes of rotation of the last two joints intersect at the center P of the moving platform. The axes of rotation of the middle two joints of each limb also intersect in points R_i equidistant to the centers of the both base and the platform [20]. The mechanism can be actuated using the first joints of any two adjacent limbs, in this case, q_{A1} and q_{B1} being the actuated ones.

The system geometry is defined by the geometric parameters α, γ and l_1, l_2, l_3. The parameter α is the angle between the middle joints for each limb, that is, axis 2/axis 3 for all the limbs. The parameter γ represents the angular offset between two adjacent limbs; in the hypothesis of equally spaced "limbs" this parameter also defines the total number of limbs in the system. The lengths l_1, l_2 and l_3 are translational offsets in the defined coordinate frames. The L-shaped link of the limb has a geometry of, $l_2 = 2l_1$. Also, l_3 can be expressed as a function of l_1 and α as $l_3 = 0.67l_1[\sin(\alpha/2) + \tan(\alpha/2)]$. Parameters l_1, l_2 and l_3, were scaled in order to obtain a unit distance from the center of the moving platform (point P) to the center of the mechanism base (point O). The parameter values of $\alpha = 45°$ and $\gamma = 90°$, which are the ones normally employed for N-UU mechanisms with 4 limbs (4-UU) [18], were chosen for this study. Given α, γ and a unit OP distance the values of l_1, l_2 and l_3 and their ratios were univocally determined.

5.6. Quaternion Joint

Figure 4f shows a CAD model for a "Quaternion joint" mechanism as proposed in [24], and based on [25]. This mechanism has a kinematic architecture of three identical limbs of two universal joints (equivalent to RRRR chains) and the joint angles q_{Lj}, $\forall L = A, B, C$ limbs and $\forall j = 1, ..., 4$ joints. This arrangement achieves a structure similar to a three-dimensional anti-parallelogram. The two universal joints of each limb are diagonally attached to the fixed base and the moving platform such that the outer axes are parallel. Also, the two axes of each universal joint have an offset.

The system geometry is fully determined with three parameters: l, o, and w. The parameter l corresponds to the diagonal distance between inner axes of the universal joints of each limb. The parameter o represents the offset between the axes of the universal joint. The parameter w signifies the radial distance of the outer joint axes from the origin of the fixed frame O. These parameters are set to be $l = 0.947$, $o = 0.056$ and $w = 0.166$. These values are proportional to the ones set by the proposing authors, but allow normalizing the size of the mechanism by setting the platform to base distance equal to 1.

It should be noted here that this mechanism proposes to only approximate the ideal spherical rolling motion, but with a very high accuracy (again see [24] for details). This error, however small, still exceeds the tolerance limits allowed by the CAD simulation tools. Thus, only for this mechanism, the authors followed an inverse kinematic approach to solve for the joint angles by minimizing the error.

5.7. Orientation Parametrization

The first four mechanisms (i.e., the 2DOF gimbal, the iCub mk.2 wrist, the five-bar mechanism and the six-bar mechanism) have an inherent gimbal-like structure. In this case, it becomes natural to choose the Roll-Pitch-Yaw Euler-angle parameterization for the platform orientations as it implies a straightforward geometric interpretation. Since the mechanisms presented in this study are 2DOF, the pitch and yaw angles were considered in the analyses while the roll for these mechanisms is always equal to zero.

Instead for the Omniwrist and the Quaternion joint mechanisms, the Tilt-and-Torsion (T & T) parameterization as proposed by Bonev et al. [34] was selected. These mechanisms fall under the class of zero-torsion mechanisms; in this case, the T&T angles yield a compact and very intuitive representation of the orientation workspace. For these mechanisms only the azimuth and tilt angles were considered in the analyses.

6. Workspace Analysis

To compare the previously presented mechanisms, the end-effector positions and orientations recorded from the CAD simulations were analyzed. The end-effector positions correspond to the Cartesian coordinates of point P with respect to the base frame attached to the fixed point O. The orientation parameterization is chosen with respect to the mechanism and is as described previously. The following subsections present the results of the CAD simulations.

6.1. Normalized Cartesian Workspace

The Figure 5 show the $\mathbb{R}^2$ plot representing the top view of the normalized Cartesian workspace for each mechanism. The plot for gimbal mechanism (Figure 5a) shows a perfect circle, signifying a full hemispherical workspace. In the case of the iCub mk.2 wrist, the hardware limitations result in a truncated section of a hemisphere (Figure 5b). The Quaternion joint mechanism also has a full hemispherical workspace (Figure 5f). However, for the other three cases, the Cartesian workspace is only a partial hemisphere. Interestingly, the top view of the workspace of the Omniwrist mechanism (Figure 5e) shows its boundaries are not symmetric with respect to the zero-abscissa and zero-ordinate axes, as reported in [30].

6.2. Orientation Angles with Regard to Joint Coordinates

The Figure 6 show the $\mathbb{R}^2 \mapsto \mathbb{R}^2$ contour mapping of the orientation angles (pitch and yaw or the azimuth and tilt angles for the respective cases) with respect to the actuator joint coordinates. These plots show the direct mapping of desired output against the input and gives a fair idea about the complexity of control law necessary for the system. A perfectly square grid for this plot, implies that the two DOF are fully decoupled, as in the case of gimbal (Figure 6a). Each of the actuator contributes exactly to 1 DOF. For the case of iCub mk.2 wrist, the yaw motion is fully decoupled whereas the pure pitch motion is dependent on both the motors (Figure 6b). For the spherical five-bars and six-bars mechanisms, the two actuators when opposite, produce a pure yaw (diagonal blue lines) and when equal produce a pure pitch (red curves). The empty spaces in the corners, result due to the failure of simulation, possibly due to singularities. Both these mechanisms achieve a very high range of motion for pitch (±90°), whereas that for yaw is restricted up to ±30° for five-bar (Figure 6c) and up to ±45° for the six-bar (Figure 6d). Both the Omniwrist and the Quaternion mechanisms (Figure 6e,f), achieve a full tilt (90°) for all values of azimuth angle ϕ. However, a peculiar "warping" (asymmetry) behavior is observed in case of the Omniwrist mechanism, as described in the previous work of the authors [30].

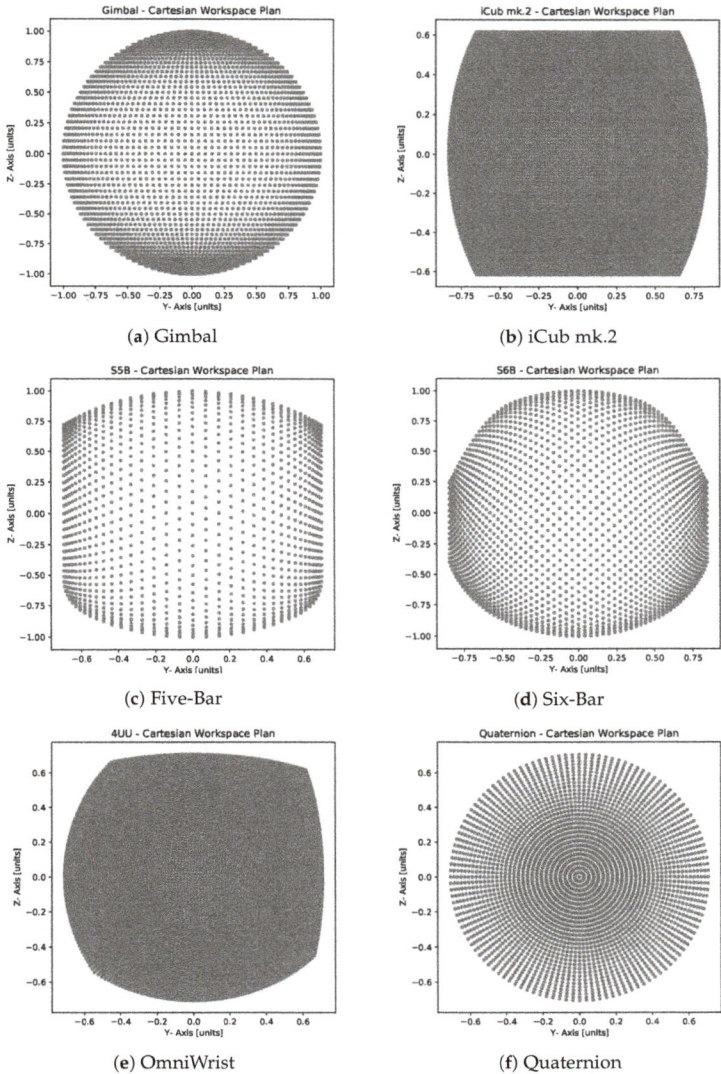

(a) Gimbal

(b) iCub mk.2

(c) Five-Bar

(d) Six-Bar

(e) OmniWrist

(f) Quaternion

Figure 5. Normalized Cartesian Workspace—Top View.

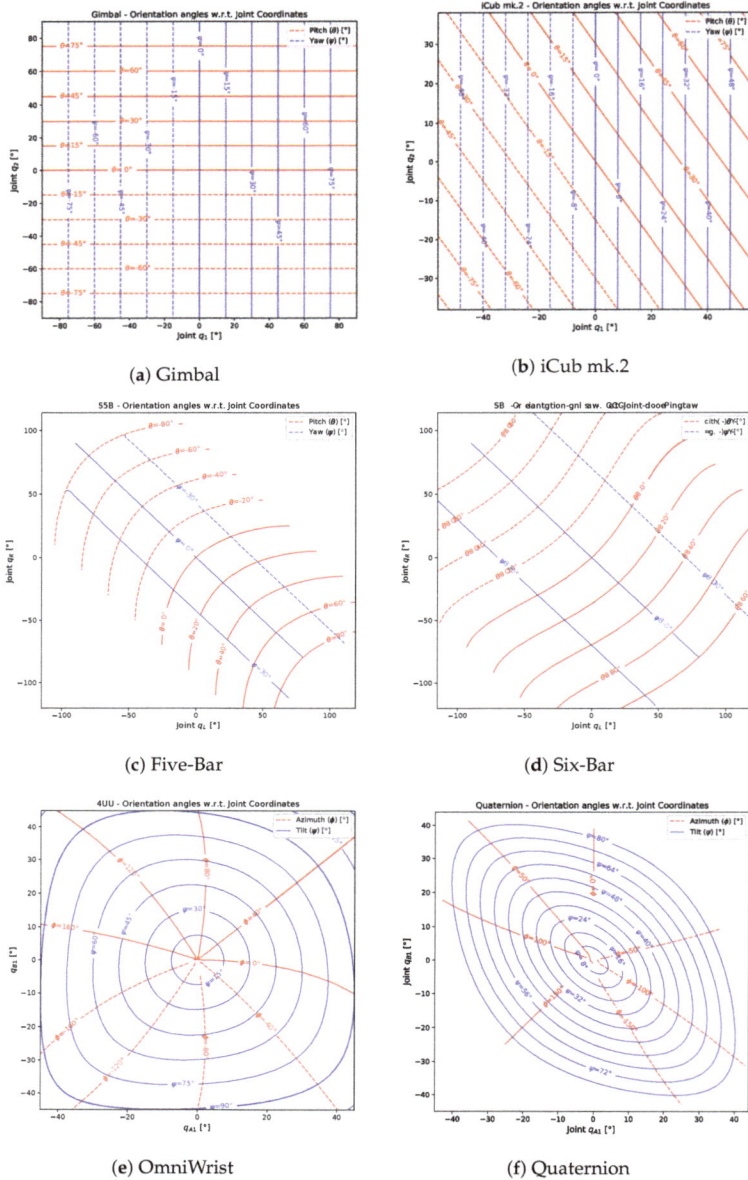

(a) Gimbal

(b) iCub mk.2

(c) Five-Bar

(d) Six-Bar

(e) OmniWrist

(f) Quaternion

Figure 6. Contour plots comparing the Orientation angles with regard to the Joint coordinates.

6.3. Orientation Angles with Regard to Normalized Cartesian Coordinates

The Figure 7 show the $\mathbb{R}^2 \mapsto \mathbb{R}^2$ contour mapping of the orientation angles (pitch and yaw or the azimuth and tilt angles for the respective cases) with respect to the normalized Cartesian coordinates of point P on the platform. These plots depict the coupling between the position and the orientation of the mobile platform. For both the spherical linkage mechanisms, only the platform yaw exhibits a linear relation with its position in the Cartesian space (Figure 7c,d) and this behavior is symmetric.

Similarly, both the Omniwrist (Figure 7e) and the Quaternion (Figure 7f) mechanisms, the platform tilts perfectly symmetric about the torsional axis (X-axis).

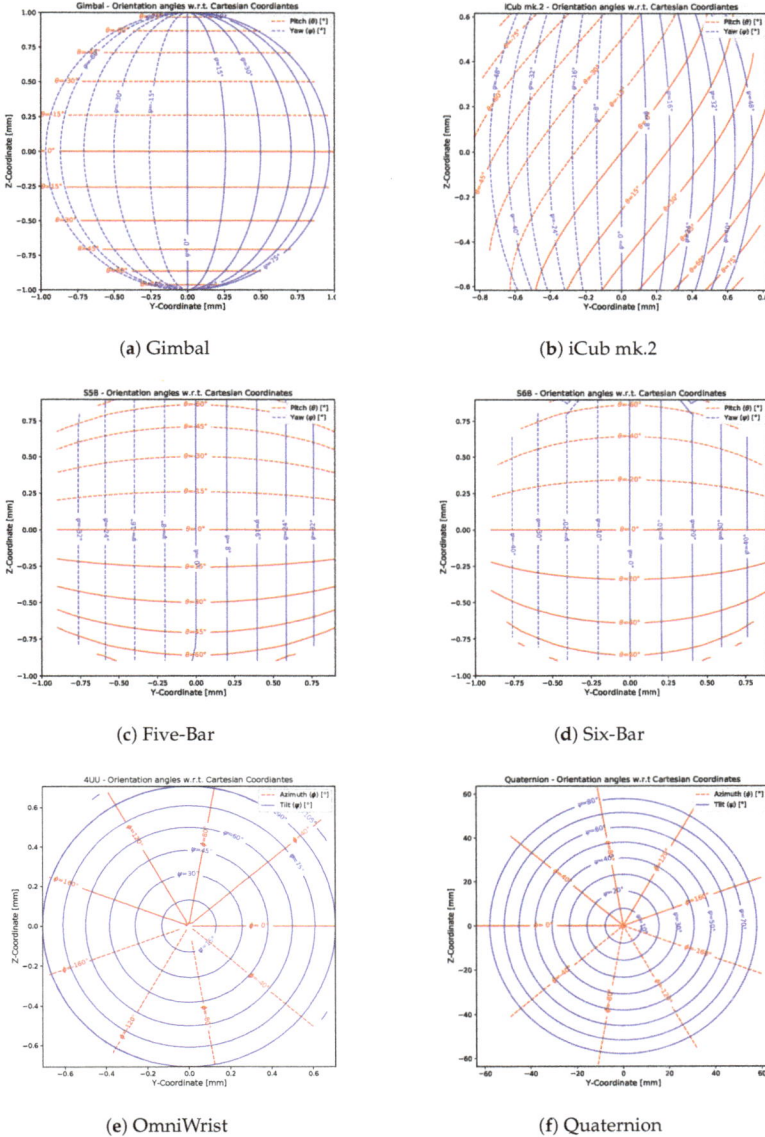

(a) Gimbal

(b) iCub mk.2

(c) Five-Bar

(d) Six-Bar

(e) OmniWrist

(f) Quaternion

Figure 7. Contour plots comparing the Orientation angles with regard to the Cartesian coordinates.

6.4. Joint Coordinates with Regard to Normalized Cartesian Coordinates

The Figure 8 show the $\mathbb{R}^2 \mapsto \mathbb{R}^2$ contour mapping of the actuator joint coordinates with respect to the normalized platform coordinates in the Cartesian space. These plots show the coupling between the platform position and the input joint angles. The plots for the gimbal (Figure 8a) and the iCub mk.2 wrist (Figure 8b) show a symmetric relation, as expected. In addition, again, it is observed that the five-bar and six-bar mechanisms (Figure 8c,d) show a quasi-linear relation albeit skewed. The Omniwrist plot (Figure 8e) is not symmetric with respect to the zero-abscissa and zero-ordinate axes, thus further implying the "warping" behavior of the workspace. On the other hand, the Quaternion joint plot (Figure 8f) is fairly regular.

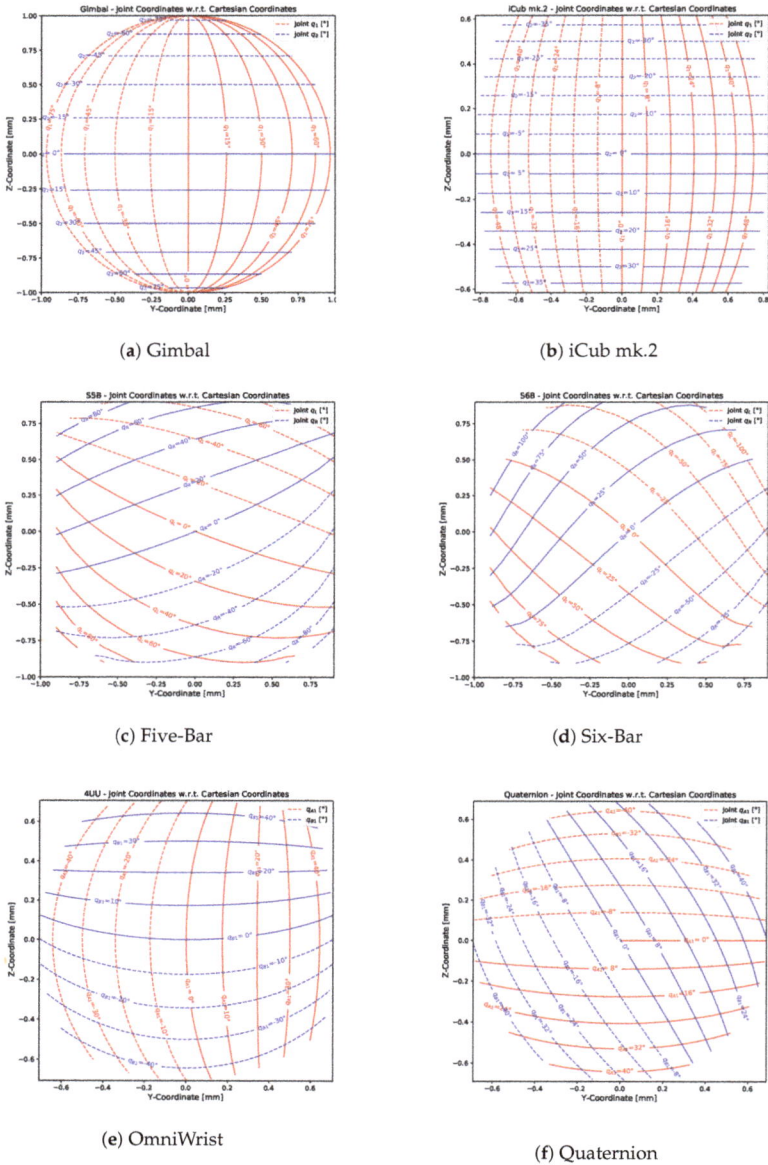

(**a**) Gimbal

(**b**) iCub mk.2

(**c**) Five-Bar

(**d**) Six-Bar

(**e**) OmniWrist

(**f**) Quaternion

Figure 8. Contour plots comparing the Joint coordinates with regard to the Cartesian coordinates.

7. Discussion

Some of the important observations from the analyses presented above are summarized in Table 2 and are discussed as follows:

- Similar to the gimbal mechanism, only the Omniwrist and the Quaternion joint mechanisms provide a full hemispherical workspace (Figure 5). Consequently, these two exhibit the highest orientational range of motion (Tilt up to 90°). The restriction of the yaw motion for the spherical linkage mechanisms arises possibly due to the presence of kinematic singularities.
- From the input to output mapping (Figure 6), only the gimbal mechanism has a perfect decoupled DOF. All the other cases show dependence on both the inputs for 1 pure DOF, except yaw motion for iCub mk.2.
- For the cases of Omniwrist and Quaternion joint mechanism, the highest amplification of inputs to the output is observed, that is, for a range of $\approx\pm45°$ of the actuators, full tilt of 90°is achieved.
- The relation between the platform position and its orientation is observed to be fairly symmetric and regular in all the cases (Figure 7).
- The relation between the input joint angles and the platform position in the case of the Omniwrist mechanism (Figure 8) illustrates an example of asymmetric "warping" behavior of the workspace.

Table 2. Mechanism Analyses Summary.

Criteria	Gimbal	iCub mk.2	Five-Bar	Six-Bar	Omniwrist	Quaternion
DOF	2	2	2	2	2	2
Decoupling	Full	Partial	Partial	Partial	None	None
Range of Motion	$\pm90°$	P $\pm56°$, Y $\pm38°$	P ±90 °; Y $\pm30°$	P $\pm90°$; Y $\pm45°$	Tilt 90°	Tilt 90°
Hemispherical Workspace	Full	Partial	Partial	Partial	Full	Full
Warping	No	No	No	No	Yes	No
Constant platform distance	Yes	Yes	Yes	Yes	Yes	No

From the analyses thus presented, both the Omniwrist and the Quaternion joint mechanisms stand out. However, the non-linear behaviors described above for the Omniwrist case have significant consequences for the actual mechanism implementation and control. Firstly, the same control input given to the system in two different configurations will yield significantly different output motions. This issue could, in theory, be solved by using configuration-dependent actuator PID gains, but this would imply a substantial complication of the existing robot control infrastructure. For these reasons, and given the desired design objectives, the Quaternion joint mechanism seems to be most suitable for implementation for the iCub humanoid wrist. Further research efforts will be devoted towards analytical kinematics and parameter optimization of this mechanism for the subsequent development of the new iCub wrist. It shall finally be noted that a series of alternative decoupled 2DOF wrist architectures were proposed by Carricato in [19]. Although simple CAD implementations of these architectures do not seem to comply with the aforementioned constraints, further work is needed to thoroughly evaluate the viability of this option for the iCub platform.

8. Conclusions

With the vision of developing a new dexterous wrist for the iCub humanoid, a comparative analysis of several state-of-the-art robot wrist implementations was presented. The spherical five-bar linkage, spherical six-bar linkage, OmniWrist-III mechanism, and the Quaternion joint mechanism were modeled and simulated using PTC Creo Parametric 4.0. The platform positions and orientation angles for each of these mechanisms were analyzed and compared against the standard 2DOF gimbal mechanism and the current iCub mk.2 wrist implementation. The Quaternion joint mechanism emerges as a promising candidate for the new iCub wrist and calls for further exploitation towards the design and development of the wrist, as well as for a better modeling of its kinematics.

Author Contributions: A.P. and D.S. conceived and designed the experiments; D.S. performed the experiments; A.P., A.S. and D.S. analyzed the data; A.P., A.S. and D.S. wrote the solver for modeling the kinematics of the Quaternion joint. G.M. contributed the analyses tools; all authors wrote the paper; Y.W. edited and revised the article.

Funding: This research received no external funding.

Acknowledgments: The authors would like to acknowledge Marco Carricato, Department of Industrial Engineering, University of Bologna; for his valuable feedback with respect to the analyses presented in this article.

Conflicts of Interest: The authors declare no conflict of interest.

References

1. Kong, X.; Gosselin, C.M. Type Synthesis of Parallel Mechanisms. In *Springer Tracts in Advanced Robotics*; Springer: Berlin, Germany, 2007; Volume 33.
2. Gogu, G. Structural synthesis of parallel robots. In *Solid Mechanics and Its Applications*; Springer: Berlin, Germany, 2009; Volume 159.
3. Fan, C.; Liu, H.; Zhang, Y. Type synthesis of 2T2R, 1T2R and 2R parallel mechanisms. *Mech. Mach. Theory* **2013**, *61*, 184–190. [CrossRef]
4. Parmiggiani, A.; Maggiali, M.; Natale, L.; Nori, F.; Schmitz, A.; Tsagarakis, N.; Victor, J.S.; Becchi, F.; Sandini, G.; Metta, G. The design of the iCub humanoid robot. *Int. J. Humanoid Rob.* **2012**, *9*, 1250027. [CrossRef]
5. Rosheim, M.E. *Robot Evolution: The Development of Anthrobotics*; John Wiley & Sons: New York, NY, USA, 1994.
6. Bajaj, N.M.; Spiers, A.J.; Dollar, A.M. State of the Art in Artificial Wrists: A Review of Prosthetic and Robotic Wrist Design. *IEEE Trans. Rob.* **2018**. [CrossRef]
7. Long, G.L.; Paul, R.P.; Fisher, W.D. The Hamilton wrist: A four-revolute-joint spherical wrist without singularities. In Proceedings of the 1989 IEEE International Conference on Robotics and Automation, Scottsdale, AZ, USA, 14–19 May 1989; pp. 902–907.
8. Farhang, K.; Zargar, Y. Design of spherical 4R mechanisms: Function generation for the entire motion cycle. *J. Mech. Des.* **1999**, *121*, 521–528. [CrossRef]
9. Yang, D.C.; Rauchfuss, J.W. A new zero-dimension robot wrist: Design and accessibility analysis. *Int. J. Rob. Res.* **2001**, *20*, 163–173. [CrossRef]
10. Stanisic, M.M.; Duta, O. Symmetrically actuated double pointing systems: The basis of singularity-free robot wrists. *IEEE Trans. Rob. Autom.* **1990**, *6*, 562–569. [CrossRef]
11. Duta, O.; Stanisic, M.M. Dextrous Spherical Robot Wrist. U.S. Patent 4878393, 1989.
12. Lemburg, J.; de Gea Fernández, J.; Eich, M.; Mronga, D.; Kampmann, P.; Vogt, A.; Aggarwal, A.; Shi, Y.; Kirchner, F. AILA-Design of an autonomous mobile dual-arm robot. In Proceedings of the Robotics and Automation (ICRA), Shanghai, China, 9–13 May 2011; pp. 5147–5153.
13. Otani, T.; Iizuka, A.; Takamoto, D.; Motohashi, H.; Kishi, T.; Kryczka, P.; Endo, N.; Jamone, L.; Hashimoto, K.; Takashima, T.; et al. New shank mechanism for humanoid robot mimicking human-like walking in horizontal and frontal plane. In Proceedings of the 2013 IEEE International Conference on Robotics and Automation (ICRA), Karlsruhe, Germany, 6–10 May 2013; pp. 667–672.
14. Kim, Y.J.; Lee, Y.; Kim, J.; Lee, J.W.; Park, K.M.; Roh, K.S.; Choi, J.Y. RoboRay hand: A highly backdrivable robotic hand with sensorless contact force measurements. In Proceedings of the 2014 IEEE International Conference on Robotics and Automation (ICRA), Hong Kong, China, 31 May–7 June 2014; pp. 6712–6718.
15. Degirmenci, A.; Hammond, F.L.; Gafford, J.B.; Walsh, C.J.; Wood, R.J.; Howe, R.D. Design and control of a parallel linkage wrist for robotic microsurgery. In Proceedings of the 2015 IEEE/RSJ International Conference on Intelligent Robots and Systems (IROS), Hamburg, Germany, 28 September–2 October 2015; pp. 222–228.
16. Ueda, K.; Yamada, H.; Ishida, H.; Hirose, S. Design of large motion range and heavy duty 2-DOF spherical parallel wrist mechanism. *J. Rob. Mechatron.* **2013**, *25*, 294–305. [CrossRef]
17. Bsili, R.; Metta, G.; Parmiggiani, A. An Evolutionary Approach for the Optimal Design of the iCub mk. 3 Parallel Wrist. In Proceedings of the IEEE-RAS 18th International Conference on Humanoid Robots (Humanoids 2018), Beijing, China, 6–9 November 2018.

18. Sofka, J.; Skormin, V.; Nikulin, V.; Nicholson, D. Omni-Wrist III- A New Generation of Pointing Devices. Part I: Laser Beam Steering Devices- Mathematical Modeling. *IEEE Trans. Aerosp. Electron. Syst.* **2006**, *42*, 718–725. [CrossRef]

19. Carricato, M. Decoupled and homokinetic transmission of rotational motion via constant-velocity joints in closed-chain orientational manipulators. *J. Mech. Rob.* **2009**, *1*, 041008. [CrossRef]

20. Wu, Y.; Carricato, M. Synthesis and Singularity Analysis of N-UU Parallel Wrists: A Symmetric Space Approach. *J. Mech. Rob.* **2017**, *9*, 051013. [CrossRef]

21. Hunt, K. Constant-velocity shaft couplings: A general theory. *J. Eng. Ind.* **1973**, *95*, 455–464. [CrossRef]

22. Wu, Y.; Löwe, H.; Carricato, M.; Li, Z. Inversion Symmetry of the Euclidean Group: Theory and Application to Robot Kinematics. *IEEE Trans. Rob.* **2016**, *32*, 312–326. [CrossRef]

23. Wu, Y.; Carricato, M. Symmetric subspace motion generators. *IEEE Trans. Rob.* **2018**. [CrossRef]

24. Kim, Y.J.; Kim, J.I.; Jang, W. Quaternion Joint: Dexterous 3-DOF Joint representing quaternion motion for high-speed safe interaction. In Proceedings of the 2018 IEEE/RSJ International Conference on Intelligent Robots and Systems (IROS), Madrid, Spain, 1–5 October 2018; pp. 935–942.

25. Lande, M.A.; David, R.J. Articulation for Manipulator Arm. U.S. Patent 4300362, 1981.

26. Townsend, W.T.; Salisbury, J.K. Mechanical design for whole-arm manipulation. In *Robots and Biological Systems: Towards a New Bionics?*; Dario, P., Sandini, G., Aebischer, P., Eds.; Springer: Berlin, Germany, 1993; pp. 153–164.

27. Seok, S.; Wang, A.; Otten, D.; Kim, S. Actuator design for high force proprioceptive control in fast legged locomotion. In Proceedings of the 2012 IEEE/RSJ International Conference on Intelligent Robots and Systems (IROS), Vilamoura, Portugal, 7–12 October 2012; pp. 1970–1975.

28. Kim, Y.J. Anthropomorphic Low-Inertia High-Stiffness Manipulator for High-Speed Safe Interaction. *IEEE Trans. Rob.* **2017**, *33*, 1358–1374. [CrossRef]

29. Ogata, M.; Hirose, S. Study on ankle mechanism for walking robots: Development of 2 DOF coupled drive ankle mechanism with wide motion range. In Proceedings of the 2004 IEEE/RSJ International Conference on Intelligent Robots and Systems (IROS), Sendai, Japan, 28 September–2 October 2004; Volume 4, pp. 3201–3206.

30. Shah, D.; Metta, G.; Parmiggiani, A. Workspace analysis and the effect of geometric parameters for parallel mechanisms of the N-UU class. In Proceedings of the International Design Engineering Technical Conferences and Computers and Information in Engineering Conference, Quebec City, QC, Canada, 26–29 August 2018; Volume 5A.

31. Shah, D.; Metta, G.; Parmiggiani, A. Comparison of Workspace Analysis for Different Spherical Parallel Mechanisms. In *IFToMM Symposium on Mechanism Design for Robots MEDER*; Gasparetto, A., Ceccarelli, M., Eds.; Mechanisms and Machine Science; Springer International Publishing: Udine, Italy, 2018; Volume 66, pp. 193–201.

32. Fiorio, L.; Scalzo, A.; Natale, L.; Metta, G.; Parmiggiani, A. A parallel kinematic mechanism for the torso of a humanoid robot: Design, construction and validation. In Proceedings of the 2017 IEEE/RSJ International Conference on Intelligent Robots and Systems (IROS), Vancouver, BC, Canada, 24–28 September 2017; pp. 681–688.

33. Sureshbabu, A.V.; Metta, G.; Parmiggiani, A. A new cost effective robot hand for the iCub humanoid. In Proceedings of the 2015 IEEE-RAS 15th International Conference on Humanoid Robots (Humanoids), Seoul, Korea, 3–5 November 2015; pp. 750–757.

34. Bonev, I.; Zlatanov, D.; Gosselin, C. Advantages of the modified Euler angles in the design and control of PKMs. In Proceedings of the 2002 Parallel Kinematic Machines International Conference, Chemnitz, Germany, 23–25 April 2002; pp. 171–188.

robotics

MDPI

Article

Development of a Novel SMA-Driven Compliant Rotary Actuator Based on a Double Helical Structure †

Rasheed Kittinanthapanya *, Yusuke Sugahara *, Daisuke Matsuura and Yukio Takeda

School of Engineering, Tokyo Institute of Technology, Tokyo 152-8552, Japan;
matsuura@mech.titech.ac.jp (D.M.); takeda@mech.titech.ac.jp (Y.T.)

*   Correspondence: rasheedo.kit@gmail.com (R.K.); sugahara@mech.titech.ac.jp (Y.S.);
Tel.: +81-80-1380-1488 (R.K.); +81-3-5734-2927 (Y.S.)

†   This paper is an extended version of our paper published in Kittinanthapanya, R.; Sugahara, Y.;
Matsuura, D.; Takeda, Y. A Novel SMA Driven Compliant Rotary Actuator Based on Double Helical
Structure. In Proceedings of the IFToMM Symposium on Mechanism Design for Robotics, Udine, Italy,
11–13 August 2018.

Received: 15 January 2019; Accepted: 13 February 2019; Published: 18 February 2019

Abstract: This paper proposes a new shape memory alloy (SMA)-driven compliant rotary actuator that can perform both passive and self-actuated motions. This SMA actuator is suitable as a redundant actuation part in a parallel robot joint to assist with singularity postures where the robot might lose the ability to maintain the position and orientation of the end effector. The double helical compliant joint (DHCJ) was chosen as a candidate mechanism; it can act in soft compliance with linear characteristics and a wide range of motion. The experimental results validated that the proposed model can be used to simulate the DHCJ behavior. The use of this mechanism exhibits advantages such as one-axis rotational motion, linear behavior even for a compliant mechanism, stiffness in the other axes of motion, and compact size. SMA leaves (strips) were used as actuation parts, and a single SMA leaf was tested before combining with the double helical frame as an SMA actuator. The prototype was fabricated, and necessary parameters such as deflection angle, temperature, torque, and stress–strain were collected to define the model for a controller. This actuator is controlled by a feedforward controller and provides rotational motion for both forward and reverse sides with a maximal range of 40 degrees.

Keywords: shape memory alloy; compliant mechanism; SMA actuator

---

1. Introduction

Shape Memory Alloy (SMA) is a special material that can recover itself to its original shape when the proper heat is applied. In the robotics field, SMA has been used as a smart actuator by providing electrical current directly to its part. Generally, the SMA part works together with a bias spring [1] because after SMA is energized, it cannot return to its initial state, therefore a bias spring would be able to bring SMA back to its initial state. Using SMA material as a smart actuator, it shows advantages such as high force per weight ratio, simple electrical current drive, and operation in silence [2]. The SMA actuator is better to apply in miniature applications than massive machines and the operation frequency is quite low compared to the other conventional actuators. SMA's actuation can be designed as translational and angular motion. Normally, the common shapes of SMA, which are produced easily, are wire-shaped, spring-shaped and strip-shaped. Farias [3] proposed a four-fingered robot hand by using SMA wire as an actuation part. Hamid [4] also designed the morphing wing mechanism by using an SMA wire and a bias spring. Yuan [5] designed the compliant actuator with SMA wire inside a plastic structure. These SMA actuators were designed for a specific applications. Wire and spring types are the most practical type for almost of SMA researches. For wire and spring types, the memorized shape is often set as a straight shape and contraction shape which both provide translational motion. If an angular or rotational motion is desired, an external mechanism is required to transform SMA movement into desired motion. Raynaerts and Van Brussel [6] proposed a technique and design aspect of shape memory alloy actuators such as a cooling system design, but this will cause the structure to become vast and massive.

To maintain the stiffness of the actuator, avoid unnecessary weight, and generalize it in a manner similar to a motor, the special compliant mechanism is considered as a frame for the actuator instead of a normal spring or other mechanism. In this study, the bias spring was replaced with a compliant mechanism called a double helical compliant joint (DHCJ) [7]. A similar concept was used with another smart material; Modler [8] designed a compliant mechanism integrated with a piezo ceramic actuator for a flap mechanism, and Hoang and Chen [9] designed a flexure parallel mechanism with the concept of selective actuation. The DHCJ consists of two materials: a hard material for the frame and a flexible material for the deformation part. This combination provides the actuator with sufficient stiffness and rotational motion. In order to move this joint, one frame is fixed to the ground and the other is pushed or pulled as a passive joint. The motion of this joint does not require any bearings or contacting parts, so this compliant joint can provide high-precision motion without lubricants, and can be washable. The purpose of this paper is to propose a new SMA compliant actuator based on the double helical structure. The characteristics of this SMA actuator were experimentally investigated, and the related parameters were determined by the experimental apparatus.

2. Design of the Double Helical Compliant Joint

2.1. Design Concept of the DHCJ

The DHCJ was originally designed by Yonemoto et al. [7] for a large-workspace compliant mechanism by using soft material for the deformation parts and rigid material for the frame. This compliant joint is a passive joint, which is suitable for application as a parallel manipulator. It consists of two similar frames with the same coordinate system, but each frame is placed 180 degrees in the z-axis from the other as shown in Figure 1a,b and connected together by leaf springs as shown in Figure 1c,d. The original DHCJ was designed to use 16 leaf springs with the overall size of 40 × 40 × 100 mm. As shown in Figure 2a, one frame is removed to show that the leaf springs are placed symmetrically at the mirror plane. There are eight pieces at the front side and eight pieces at the back side as shown in Figure 2b. This kind of arrangement can constrain undesired motion from the other five axes so that the joint rotates only around the z-axis. Accordingly, the joint can be estimated as working as a one degree-of-freedom mechanism [10].

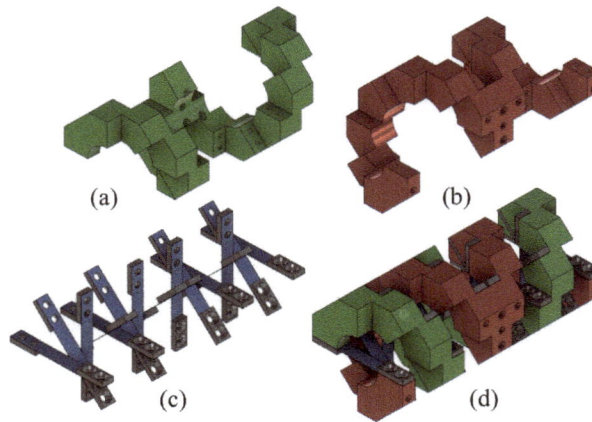

Figure 1. Components of a double helical compliant joint (DHCJ): (**a**) First frame. (**b**) Second frame. (**c**) Leaf springs. (**d**) Assembly of all parts.

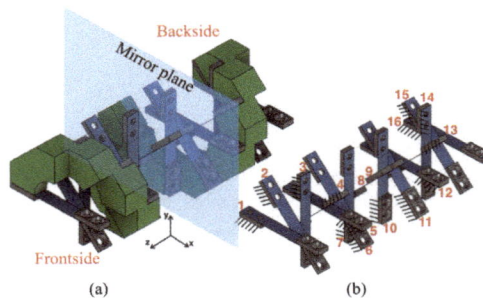

Figure 2. Half of a DHCJ: (**a**) One helical frame with mirror plane. (**b**) Arrangement of leaf springs.

2.2. Modeling of the DHCJ

In order to model the DHCJ, the chained-beam constraint model (CBCM) was used because it is simple to use in calculation and simulation of the result. The CBCM is a numerical method which developed by Ma and Chen [11] to be used in a compliant mechanism with various constraint conditions. The DHCJ can be considered as 16 elastic beams fixed at one end and constrained at the other end to move in a circular path. The same load was applied to each elastic beam, so only just one model multiplied by 16 would yield the whole stiffness of the mechanism, as shown in Figure 3. The experiment were performed to verify that the proposed method can predict the actual behavior of the DHCJ. The apparatus is shown in Figure 4, where the DHCJ is driven from an initial angle to the maximum angle. The results from Figure 5 show that the DHCJ acts as a soft linear compliant mechanism, which is suitable for adoption with other mechanisms as a rotational motion component. From the characteristics graphs, the double helical structure with proper arrangement of the leaf springs can transform the nonlinear system mechanism (large deflection of an elastic beam) to become simply a spring-like mechanism.

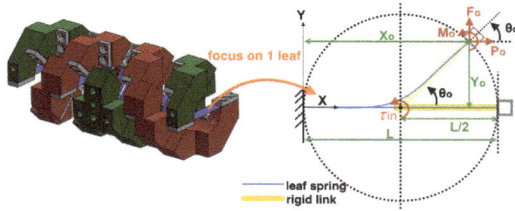

Figure 3. Kinematics model of one leaf spring from the DHCJ.

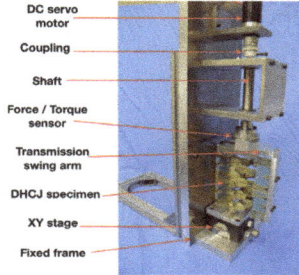

Figure 4. Characteristics testing apparatus.

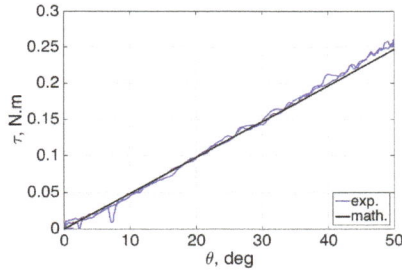

Figure 5. Comparison of simulation and experimental results.

3. Design of the SMA Actuator

This section describes the design of the SMA component and the DHCJ in combination as a special compliant SMA actuator. The idea is to keep the size compact, by replacing some leaf springs of the DHCJ with SMA, and without increasing the external mass of the system. When no power is applied to the system, the actuator behaves as a passive compliant joint. When the proper energy in applied to it, the actuator transforms from a passive to an active component. The modeling and the idea of this conceptual design are shown.

3.1. Modeling of the SMA Leaf

SMA is a special material whose phase changes when the proper heat is applied to it. At the low-temperature or martensite phase (M), the crystalline structure of an SMA is quite flexible and can be easily deformed into new shapes when force is applied. Upon heating, the crystalline structure returns to a memorized configuration in austenite phase (A). The transformation of A to M and M to A are characterized by four temperatures: A_s and A_f are the start and finish temperatures, respectively, for the A phase; M_s and M_f are the start and finish temperature, respectively, for the M phase. The phase transformation curve can be expressed in terms of a martensite fraction, ξ. The martensite fraction is bounded between 0 (fully austenite) and 1 (fully martensite) as shown in Figure 6. As seen in the phase transformation, SMAs show significant hysteresis during the two transformations. Many studies show that the behavior of SMA depends not only on temperature, but also on the stress acting on it. At any value of stress or temperature, the material can be in one of three states: fully martensite, fully austenite, or a mixture of austenite and martensite. The phase transformation of the SMA can be expressed as

$$\xi_{M \to A} = 0.5[cos(a_A(T - A_s) - \frac{a_A}{C_A}\sigma) + 1]$$
$$\xi_{A \to M} = 0.5[cos(a_M(T - M_f) - \frac{a_M}{C_M}\sigma) + 1],$$

(1)

where $a_A = \pi/(A_f - A_s)$ and $a_M = \pi/(M_s - M_f)$, are material constants. T is the applied temperature, σ is the pre-stress, and C_A and C_M are the slope of the stress and transformation temperature from experimental data.

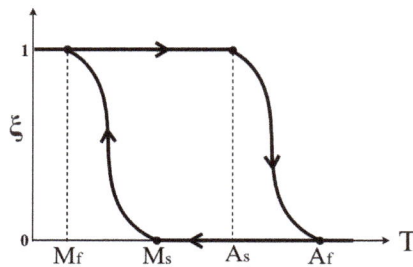

Figure 6. Phase transformation of shape memory alloy (SMA) as a function of temperature.

For simplicity, for the first prototype, the one-dimensional constitutive model from Liang and Roger [12] was used to describe the SMA behavior as shown in Equation (2).

$$\sigma - \sigma_0 = E(\epsilon - \epsilon_0) + \Omega(\xi - \xi_0),$$

(2)

where σ and σ_0 represent the stress and the pre-stress of the SMA actuator; E is the Young's modulus and ϵ and ϵ_0 are the current and initial strain; ξ and ξ_0 are the current and initial value of martensite fraction and Ω is the transformation coefficient of the material constant. The coefficient can be obtained from the residual strain and modulus of elasticity as $\Omega = -E\,\epsilon_L$. From this equation, when SMA changes its phase from martensite to austenite, the strain changes because of the SMA motion. The change from the right-hand side of the equation causes the SMA to generate stress which turn induces force and torque to the system. In this research, the SMA strip type was designed in a similar shape as the leaf spring in DHCJ as shown in Figure 7. The memorized shape (austenite phase shape) was set as a U-shape as shown in Figure 8. The material composition of the SMA using in this research is NiTi-Cu with 49.2% of Ni, 5.8% of Cu and the remaining of Ti. Considering a single leaf, with one end was fixed and the other end was bent in a circular path, the forward kinematics transforms the deflection angle θ to strain ϵ by considering the arc shape as shown in Figure 9. When the SMA leaf bends into each angle, the deflection angle θ is equal to arc angle ψ, so the strain can be expressed in term of deflection angle as

$$\epsilon = \frac{-\psi(r - \frac{t}{2}) - l}{l},$$

(3)

where t is the leaf's thickness, l is the leaf's length, ψ is the arc angle (where $\psi = \theta$) and r is the radius of curvature, where $r = (l/2)/tan(\psi/2)$.

In the robotics field, the proper way to energize the SMA is to use electrical power, Ramio and Edwin [2] proposed a classical differential equation that expresses the heat store in SMA as

$$\rho C V \frac{dT}{dt} = i^2 R - h A (T - T_e) \tag{4}$$

where ρ is the density of the SMA, C is the specific heat, V is the volume of the SMA, i is the electrical current, R is the elements electrical resistance, h is the heat-exchange coefficient between the SMA and the surrounding air, A is the surface area of the SMA, T_e is the environmental temperature and T is the SMA temperature at time t.

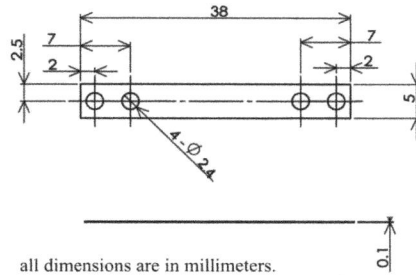

all dimensions are in millimeters.

Figure 7. Dimensions of SMA leaf.

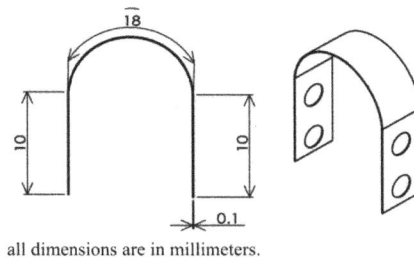

all dimensions are in millimeters.

Figure 8. SMA's memorized shape.

When electrical current is applied to the SMA leaf, it can generate a bending moment, which can be considered as generated torque τ_g from the SMA. The equation of rotational motion can be expressed as

$$\tau_g = J\ddot{\theta} + D\dot{\theta} + k\theta, \tag{5}$$

where J is the moment of inertia, D is the damping coefficient, k is the rotational stiffness and θ is the deflection angle.

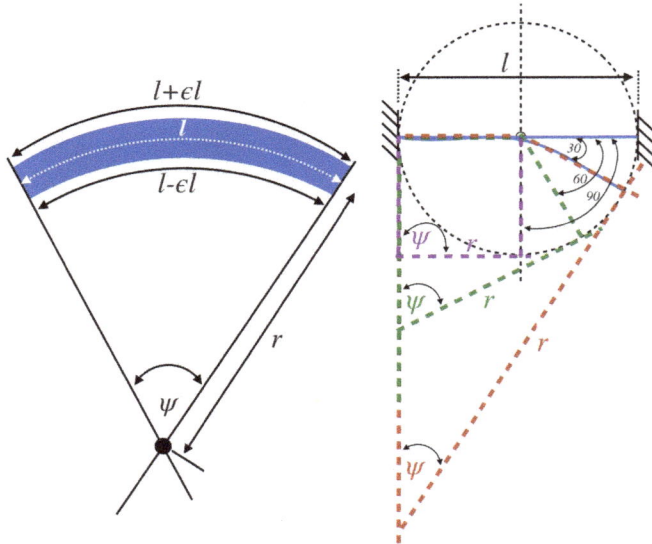

Figure 9. Approximated strain from arc shape of the beam.

By combining each elements as proposed previously, the overall block diagram of the SMA driven compliant actuator is as shown in Figure 10. First, the proper voltage was applied to the SMA strip; then from the heat transfer model of Equation (4), the heat changes the crystalline structure of the SMA from martensite (initial shape) to austenite state (memorized shape) as expressed in Equation (1). Those changes induced the SMA to generate stress or torque on the system as expressed in Equation (2). Because of the input torque, the mechanical system generated motion according to the dynamic parameters and yielded angular motion as the output. From this concept block diagram, the feed-forward controller is considered to evaluate the precision of the model from theory to practice.

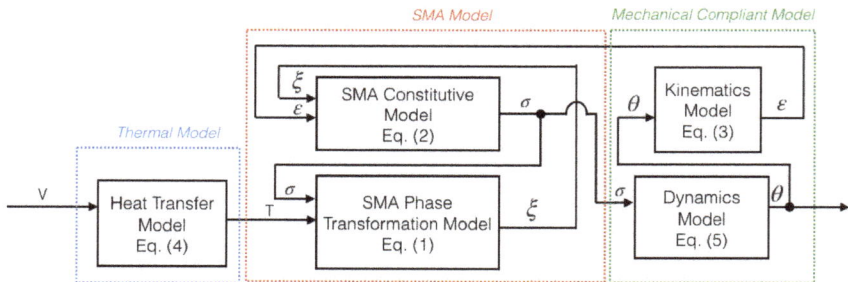

Figure 10. Block diagram of the SMA actuator system.

3.2. Characteristics of a Single SMA Leaf

In order to understand the physical behavior of the SMA before combining it with the DHCJ frame, a single SMA leaf was tested to validate the model. A driver circuit and proper mechanical structure were also tested to validate the concept. The approximated kinematics of DHCJ with the SMA leaf and leaf spring are shown in Figure 11. According to the DHCJ motion that can rotate in a circular path, the rigid link that acts as the radius arm is attached at the upper and lower parts. Two bearings are used to constrain the motion only around the principal axis. There are two SMA leaves for forward and reverse driving. A pre-bent angle was added to both leaves to avoid the beam buckling shape problem. When the SMA is energized at the forward side, the reverse SMA is bent to a straight shaped. Similarly, when the reverse SMA is energized, the forward SMA will become straight either as shown in Figure 12.

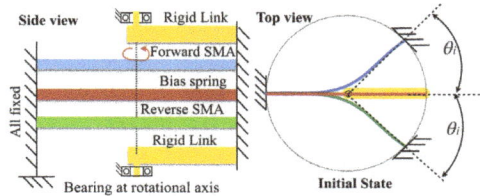

Figure 11. Estimated kinematics of DHCJ with an SMA leaf.

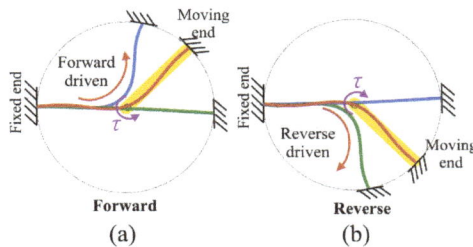

Figure 12. Energizing the SMA. (**a**) Forward drive, (**b**) reverse drive.

A pulse-width modulation (PWM) drive was used because it is an efficient way to control the SMA and it exhibits a linear proportional relationship between the power and duty cycle. To provide the proper amount of electrical energy, a high drain lithium ion battery (IMR 18650) was used. To generate heat in a short time, the SMA strip was shorted-circuited. For this experiment, the circuit shown in Figure 13 was used as the SMA driver. The various pre-bent angles of SMA as shown in Figure 14 were tested with the same configuration, and the result shows that from no pre-bent angle to a small pre-bent angle, the SMA seemed to give much power to drive, but in fact the opposite SMA resisted the motion, resulting in only a small angle movement. A large pre-bent angle, such as 45 degrees, yielded the expected result in which the overall motion turned by only a small angle because the initial point was already half of the memorized shape, so the power was not enough to induce a large deflection angle. The pre-bent angle of 30 degrees was chosen because it can have the largest deflection angle at 30 degrees for each side.

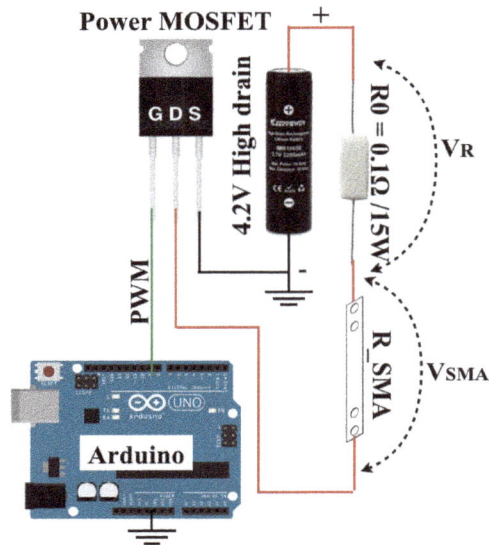

Figure 13. SMA drive circuit.

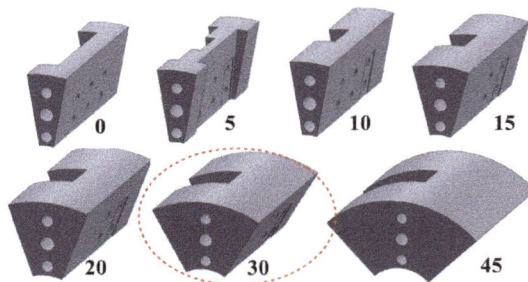

Figure 14. Different pre-bent angle of SMA leaf.

The ambient temperature during testing was 20–22 °C. The SMA started to rotate when it reached the temperature of 38–40 °C, and reached the maximum angle of 30 degrees at a temperature of approximately 50–55 °C. When the power was released, the temperature decreased to approximately 33 °C while maintaining the position at the same angle; then the bias spring tried to rotate the SMA back to its initial position. The deflection angle and temperature of the SMA are represented with respect to time by changing the duty cycle to 45%, 60% and 75% as shown in Figure 15. A 45% duty cycle yielded a lower temperature, but it required more time to reach the maximum angle. Figure 15a shows that the higher voltage can decrease the SMA response time, but it also caused higher temperature. When the energy was released during the "off" time, all of the SMA's temperature dropped down similarly from the maximum temperature to approximately 30 °C, then linearly decreased to room temperature as shown Figure 15b. The testing results show that the SMA strip works from 45% to 75% duty cycle with 4 V amplitude and 10 A drawing currents; for lower duty cycles, the SMA only slightly changed and never reached the maximum angle of 30 °C.

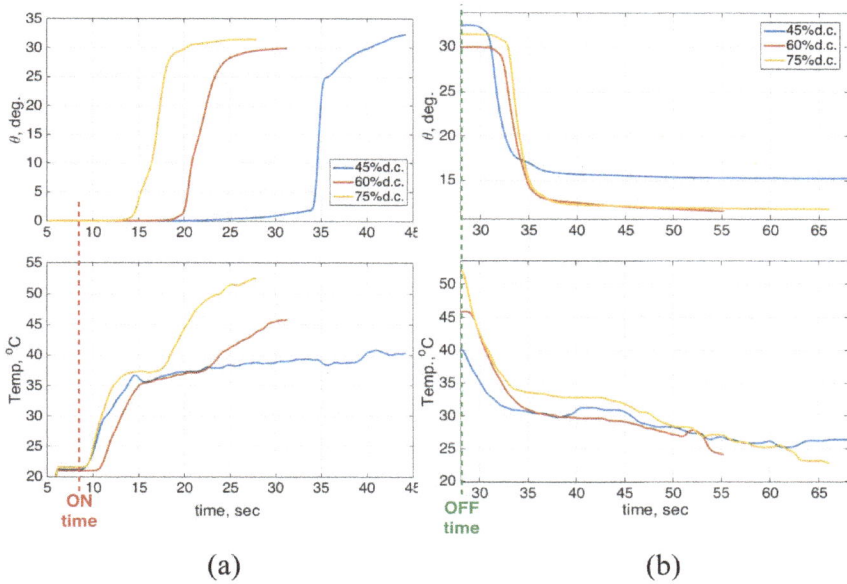

Figure 15. Deflection angle, temperature with respect to time. (**a**) "On" period. (**b**) "Off" period.

3.3. An Actuator Design

The information from the test described previously was used to design the combination of SMA strip and double helical structure. The SMA leaves were divided into four pieces for the forward SMA and four pieces for the reverse SMA, and all of the SMA components were pre-bent to some angle. The leaf springs were chosen as eight pieces, with a straight shape and half of the width in size to make the actuator soft in the z-axis rotation and stiff in the other axes, as shown in Figure 16a,b. The special double helical frame for the SMA actuator was designed according to the requirement of the SMA strip attachment. The size of the actuator was 40×40 mm for a cross-section size with 140 mm long, and total weight was 95 grams. The leaf arrangement is shown in Figure 16c; red indicates the bias springs, blue indicates the SMA forward driving leaves, and green indicates SMA backward driving leaves. When the forward SMA was energized to return to the memorized shape, the reverse SMA acted as a bias spring and deflected to the straight shape as described in the conceptual design presented in Section 3.2 ; energizing the reverse SMA case is similar.

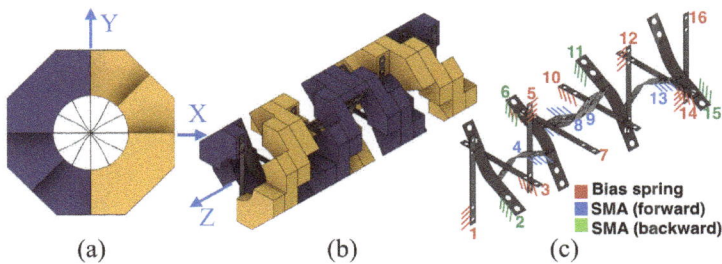

Figure 16. CAD model of the SMA actuator. (**a**) Front view. (**b**) Perspective view. (**c**) Arrangement of bias springs and SMA leaves.

4. Experiments

4.1. An Actuator Test

In order to model the actuator and compare it with the experimental result, some parameters were determined by experiments. Dynamic parameters, including rotational stiffness, damping coefficient and inertia were found by using the apparatus in Figure 4. From the test, the relationship between reaction torque and angle is shown in Figure 17. The graph shows significantly that the low rotational stiffness is almost linear. According to that, the mechanism of this SMA actuator can be approximated as a soft linear spring with a constant rotational stiffness (k = 0.027 N·m/rad). A simple free vibration test was performed to determine the values of damping ratio and natural frequency. From the results of the free vibration test, the logarithmic decrement method can be applied, and the value of damping coefficient and inertia were solved as 4.6787×10^{-4} N·s/m and 3.207×10^{-4} kg·m^2, respectively.

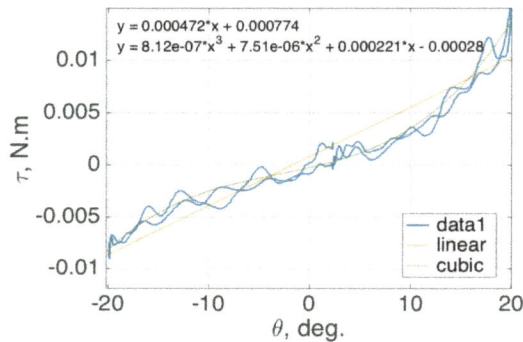

Figure 17. Rotational stiffness curve of the SMA actuator.

We considered the voltage as the input of an actuator and the angle as the output. Thus, to understand how this SMA actuator responds to a step input, the duty cycle was changed from 1% to 75% with an increment of, and 4 V amplitude was used for the PWM signal. Figure 18 shows the SMA actuator prototype with testing apparatus and connecting wire. The maximum output angle and the input duty cycle are plotted as shown in Figure 19 to show the response of the actuator. For input lower than 15%, the actuator moves slightly, only 0.2 degrees, but in the range of 15% to 22%, the output angle was significantly changed. From 22% to 45%, the output angle continued to increase, but over 45%, the output angle remained steady at a maximum value 22 degrees. Using the apparatus in Figure 18, the SMA actuator was energized with different input values and the responding angle and the temperature were measured. The results are plotted in Figure 20. When the energy was removed from the SMA, the actuator returned rapidly while temperature decreased slowly. This caused a large gap between the heating and cooling path as a hysteresis loop. The austenite start temperature, A_s, is considered as the temperature to make an actuator start to rotate, and the austenite finish temperature, A_f is considered as the temperature at which the actuator remains steady at the final value, as shown in Figure 21. These temperatures are defined as 30 °C and 50 °C for A_s and A_f, respectively. The martensite start and finish temperature were used in the cooling path model, but our scope focused on the heating path used to energize the SMA and considered the cooling path as ambient cooldown so there was no need to determine the M_s and M_f temperatures.

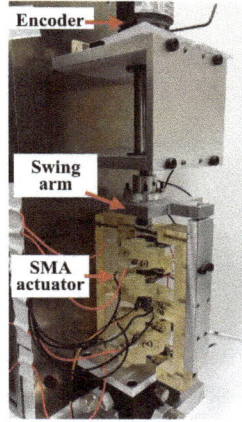

Figure 18. Apparatus for testing the SMA actuator.

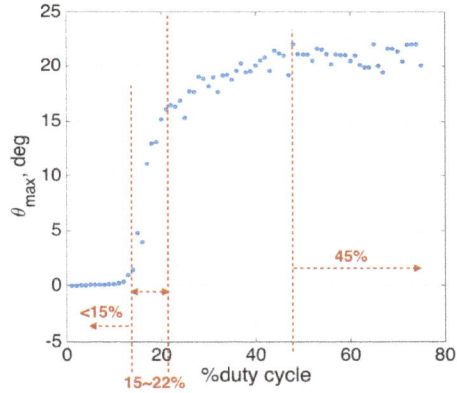

Figure 19. Input and output relation.

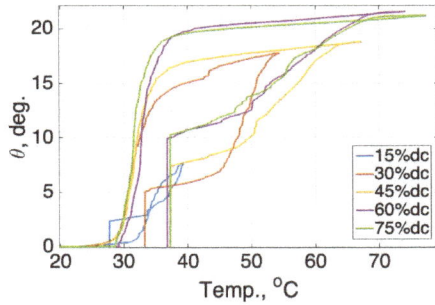

Figure 20. Hysteresis loop of angle and temperature.

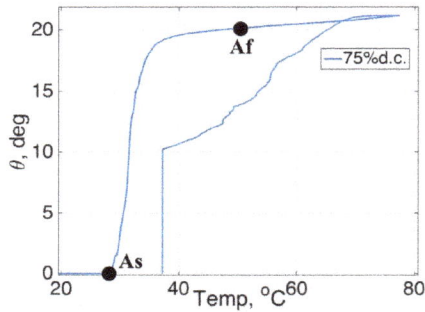

Figure 21. Determination of the austenite/martensite temperature.

The generated torque was tested by using the apparatus shown in Figure 4; the actuator was the double helical compliant SMA, and a motor was used to lock the shaft of the mechanism, so the force/torque sensor could measure the responding torque due to the PWM input power. The torque and temperature with respect to time are shown in Figure 22. The relationship of torque and temperature is plotted as shown in Figure 23. The hysteresis loop of deflection angle with respect to the temperature is shown in Figure 20.

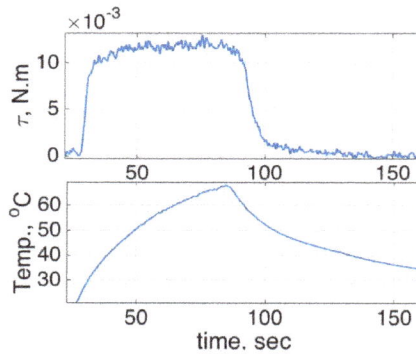

Figure 22. Generated torque and temperature with respect to time.

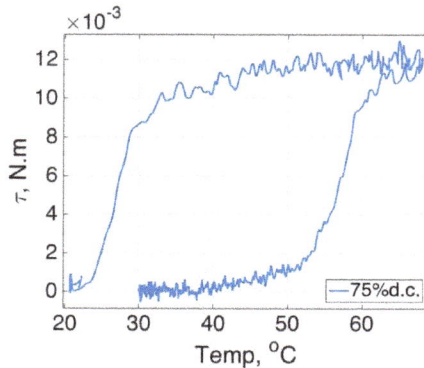

Figure 23. Generated torque with respect to temperature.

From these experiments, all of the data were collected to calculate other parameters such as C_A and C_M as mentioned in Equation (1), which were calculated as $C_A = 188.825$ MPa/°C, $C_M = 156.8$ MPa/°C. Residual strain $\epsilon_L = 0.005376$ for calculating the transformation coefficient $\Omega = -E\,\epsilon_L$.

4.2. Comparison with the Model

The model from Figure 8 was simulated with a step input of 75% duty cycle with a 60 s duration. The torque and angle of the simulation and experimental results were compared as shown in Figure 24. The simulation result of angle was slightly lower than the actual behavior, but the torque was quite similar with the maximum at 12 N·mm. Figure 25 shows that the higher input voltage helped the system to respond faster. For 60 and 75% duty cycle, the output angle from experimental result remains steady approximately 21.16 degrees and 21.55 degrees with an error of around 0.5 degrees from the simulation results. In the case of 15% duty cycle, the output angle from the experimental results remained steady at approximately 18.4 degrees with an error of approximately 0.5 degrees from the simulation result.

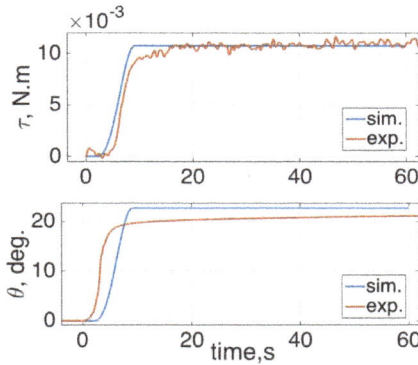

Figure 24. Comparison of torque and deflection angle from experimental and simulation results of the SMA actuator.

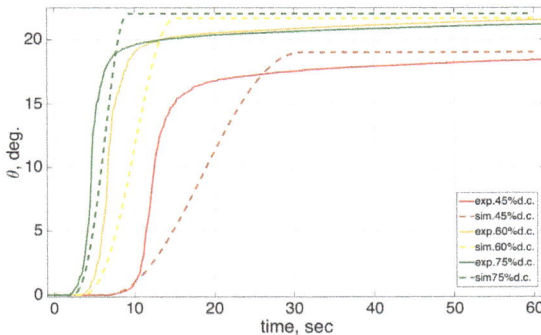

Figure 25. Comparison of a deflection angle with a different input.

The model can be applied to the feed-forward controller by inverting the model and placing it before the first block of Figure 10, the input to the overall block now is the desired angle, but as Figure 19 shows, the controllable range of the actuator was from 16 degrees to 22 degrees. The result from experiment and simulation as shown in Figure 26 indicates that the actual behavior of this SMA actuator responded faster than the simulation result, but for higher desired angles, both results responded similarly. Figure 27 shows top and side views of the apparatus when driving the SMA actuator was driven. The green dashed indicator shows the driving angle from the actuator, and the red dashed indicator shows a neutral position of the SMA actuator. Figure 28 shows top and side views of the apparatus during the return motion. To reach the maximum angle from the neutral position, the SMA actuator used a short time of approximately five seconds. However, because this SMA actuator used ambient air for a cooling system, the return motion from the maximum angle to neutral position took approximately 18 s. However, this problem can be solved by driving the reverse SMA to make it return faster. From the single SMA leaf testing in Section 3 and the SMA actuator testing in Section 4, the maximum driving angle is approximately 20 degrees which lower than the pre-bent angle of SMA attachment, so when the SMA actuator fully turns, the bias SMA wouldn't be driven to straight shape and still remain a space for a higher angle. Accordingly, leaf buckling wouldn't occur.

One possible solutions to make this actuator more controllable is to adapt the feedback control to compensate for the steady-state error. The self-sensing method is thus an interesting method to apply with this SMA actuator and the ability of a compliant mechanism still remains. The other solution is to develop the 3D constitutive model with the exact equations of stress and strain distribution, and apply the complete SMA model to feedforward control again.

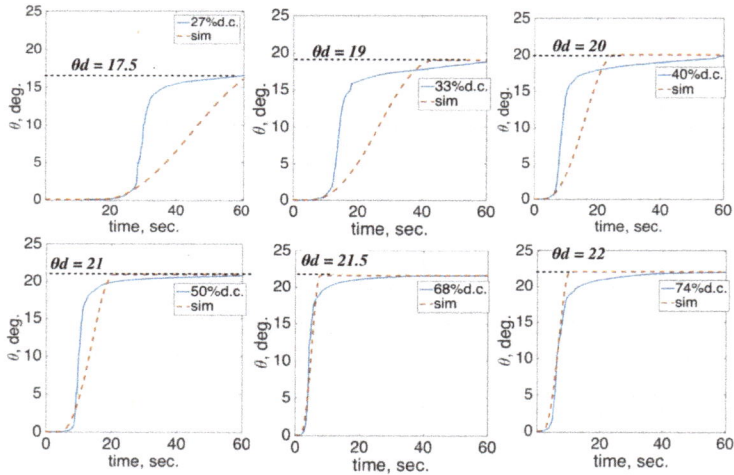

Figure 26. Applying the SMA model as feed-forward control.

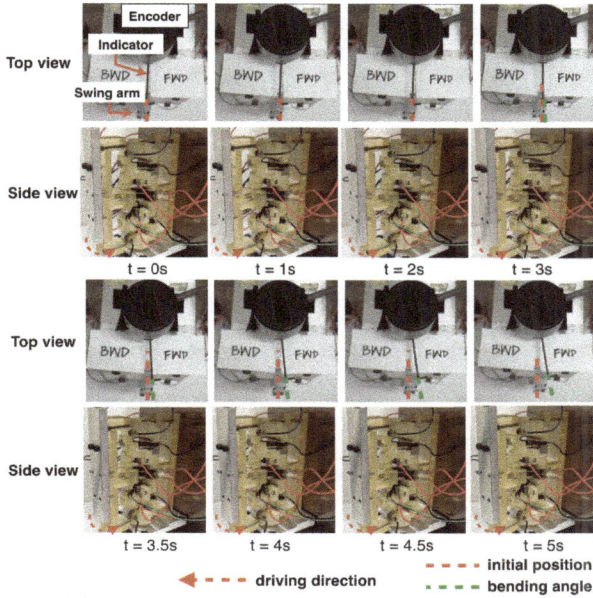

Figure 27. Driving motion of the SMA actuator from top view and side view.

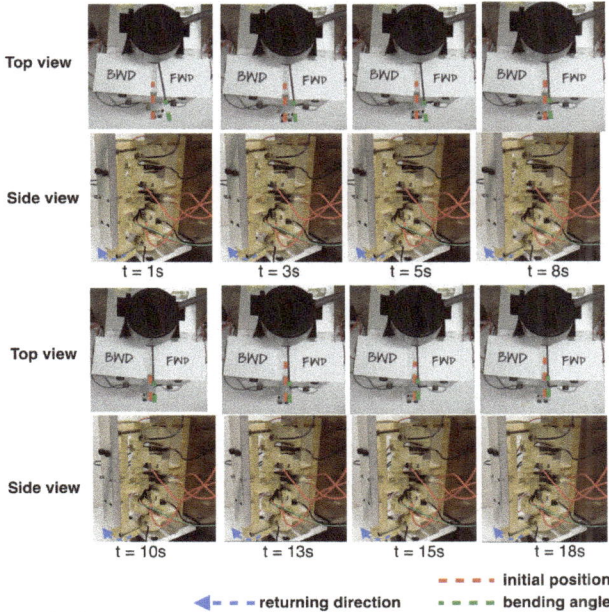

Figure 28. Returning motion of the SMA actuator from top view and side view.

5. Conclusions

This paper proposes the design of a new SMA-driven compliant actuator based on a double helical structure. The DHCJ characteristics of the testing and simulation results agree that using this mechanism can act as a soft compliant mechanism with a linear spring behavior. Testing a single SMA leaf provides information about the input power, maximum angle and a proper pre-bent angle. The combination of leaf springs and the SMAs was symmetrically designed to provide a spring-like behavior, and can act as a passive or active component. From the testing of the prototype, some necessary parameters in the model were determined by the experiments. This SMA actuator can be essentially controlled by using the feed-forward controller from the SMA model. This new SMA-driven compliant actuator has a working range of $+/-$ 22 degrees and a generated torque of 12 N·mm.

For future work, a 3D model of the SMA must be used to model the feedforward controller and a feedback signal such as the current temperature or the deflection angle should be implemented together to make the actuator more controllable and yield more accurate results. As an application, this SMA actuator is considered for implementation on a parallel robot in order to check the performance of the compensated motion when the robot is in various singularity postures. To overcome the slow actuation of the SMA, a cooling technique is also considered for a faster response operation.

Author Contributions: Conceptualization, R.K. and Y.S.; methodology, R.K.; software, R.K., Y.S. and D.M.; validation, Y.S., D.M. and Y.T.; writing–original draft preparation, R.K.; writing-review and editing, R.K., Y.S. and Y.T.; supervision, Y.S. and Y.T.

Funding: This work was partially supported by JSPS KAKENHI Grant Number JP17H03162, and JKA and its promotional funds from the AUTORACE.

Conflicts of Interest: The authors declare no conflict of interest.

Abbreviations

The following abbreviations are used in this manuscript:

DHCJ   Double Helical Compliant Joint
SMA   Shape Memory Alloy
CBCM   Chained-Beam Constraint Model

References

1. Jani, J.M.; Leary, M.; Subic, A. Designing shape memory alloy linear actuators: A review. *J. Intell. Mater. Syst. Struct.* **2017**, *28*, 1699–1718. [CrossRef]
2. Ramio, V.; Edwige, E.P. Modeling and Temperature Control of Shape Memory Alloys With Fast Electrical Heating. *Int. J. Mech. Control* **2012**, *13*, 2.
3. Farias, V.; Solis, L.; Melendez, L.; Garcia, C.; Velazquez, R. A four-fingered robot hand with shape memory alloys. In Proceedings of the AFRICON 2009, Nairobi, Kenya, 23–25 September 2009.
4. Hamid, B.; Aghil, Y.K.; Mohammad, R.Z.; Seyed, S.M. Experimental study of a bio-inspired robotic morphing wing mechanism actuated by shape memory alloy wires. *Mechatronics* **2014**, *24*, 1231–1241.
5. Yuan, H.; Balandraud, X.; Fauroux, J.C.; Chapelle, F. Compliant Rotary Actuator Driven by Shape Memory Alloy. *Mech. Trans. Robot. Mech. Machine Sci.* **2017**, *46*, 343–350.
6. Reynaerts, D.; Van Brussel, H. Design aspects of shape memory actuators. *Mechatronics* **1998**, *8*, 635–656. [CrossRef]
7. Yonemoto, K.; Takeda, Y.; Tong, Z.; Higuchi, M. A New Flexure Revolute Joint with Leaf springs and Its Application to Large Workspace Parallel Robot. *J. Adv. Mech. Design Syst. Manuf.* **2012**, *6*, 76–87. [CrossRef]
8. Modler, N.; Modler, K.H.; Hufenbach, W.; Lovasz, E.C.; Perju, D.; Margineanu, D. A Design of Compliant Mechanism with Integrated Actuators. In Proceedings of the 10th IFToMM International Symposium on Science of Mechanisms and Machines, Brasov, Romania, 12–15 October 2009; pp. 655–664.
9. Hoang, P.; Chen, I.M.; Yeh, H.C. Micro-Manipulation System Design Based On Selective-Actuation Mechanisms. *Int. J. Rob. Res.* **2006**, *25*, 171–186.
10. Kittinanthapanya, R.; Sugahara, Y.; Matsuura, D.; Takeda, Y. Modeling and Char- acterization of the Double Helical Compliant Joint. In Proceedings of the International Symposium on Robotics & Mechatronics 2017, Sydney, Australia, 29 November–1 December 2017.

Robotics **2019**, *8*, 12

11. Ma, F.; Chen, G. Modeling Large Planar Deflections of Flexible Beams in Compliant Mechanisms Using Chained Beam-Constraint-Model. *J. Mech. Robot.* **2016**, *8*, 021018. [CrossRef]
12. Liang, C.; Roger, C.A. One-dimensional Thermomechanical Constitutive Relations for Shape Memory Alloy Material. *J. Intell. Mater. Struct.* **1997**, *8*, 285–302. [CrossRef]

robotics

MDPI

Article

Use of McKibben Muscle in a Haptic Interface †

Walter Franco *, Daniela Maffiodo, Carlo De Benedictis and Carlo Ferraresi

Department of Mechanical and Aerospace Engineering, Politecnico di Torino, 10129 Torino, Italy;
daniela.maffiodo@polito.it (D.M.); carlo.debenedictis@polito.it (C.D.B.); carlo.ferraresi@polito.it (C.F.)
* Correspondence: walter.franco@polito.it; Tel.: +39-011-090-3348
† This paper is an extended version of our paper published in Franco, W.; Daniela, M.; Benedictis, C.D.;
 Ferraresi, C. Dynamic Modeling and Experimental Validation of a Haptic Finger Based on a McKibben Muscle.
 In Proceedings of the IFToMM Symposium on Mechanism Design for Robotics, Udine, Italy, 11–13
 August 2018.

Received: 17 January 2019; Accepted: 14 February 2019; Published: 18 February 2019

Abstract: One of the most relevant issues in the development of a haptic interface is the choice of the actuators that are devoted to generating the reflection forces. This work has been particularly focused on the employment of the McKibben muscle to this aim. A prototype of one finger has been realized that is intended to be part of a haptic glove, and is based on an articulated mechanism driven by a McKibben muscle. A dynamic model of the finger has been created and validated; then, it has been used to define the control algorithm of the device. Experimental tests highlighted the static and dynamic effectiveness of the device and proved that a McKibben muscle can be appropriately used in such an application.

Keywords: pneumatic artificial muscle; McKibben muscle; haptic glove; hand exoskeleton; teleoperation; force reflection; human-machine interaction

1. Introduction

Teleoperation consists of the control of a remote machine or device (slave), using an appropriate interface, called a master. The master is haptic if it is able to transmit the sense of touch, for example generating a force reflection to the operator. In general, haptics refers to the study of touch feedback in different applications, including virtual reality (such as for example in medical simulators), gaming, and tele-robots (such as surgical robots or remote manipulators for space exploration).

Among the different haptic devices that are able to generate haptic feedback on different sites of human body, the haptic glove is the most used, because hands play a dominant role in perception and manipulation tasks when grasping virtual objects [1].

When designing a haptic glove, the architecture of the mechanism that is able to transmit the force to the single fingers, as well as the type of sensors and the actuation, must be chosen. Generally, the problem has to be solved for a single finger, and replicated to all the fingers of a hand.

Starting from the performance specifications in terms of degrees of freedom, weight, size, dexterous capabilities, and safety level, multiple mechanisms for the transmission of the reflection force have been developed [2]. In particular, in order to assure the coincidence between the center of the relative rotation of the links of the mechanism and the mean physiological rotational axes of the joints of the fingers, different solutions have been designed. Heo et al. [2] classified five mechanical architectures: serial linkages with rotational joints that are coincident with the physiological rotational axes, linkages with a remote center of rotation, redundant linkages [3], tendon-driven mechanisms, and serial linkages attached to distal segment.

As regards the sensing of the user's intended motion, the reflection force on the finger and the position of the joints should generally be measured. Heo et al. [2] described some different solutions that have been adopted in the literature. For the measurement of the contact force, force sensing resistors, pneumatic pressure sensors, and strain gauge sensors are frequently used. For the measurement of motion sensing, a bending sensor or a rotary encoder can be used.

To drive the mechanism that is devoted to transmitting the reflection force to the finger, different types of actuators have been used. Conventional electric motors have been efficaciously employed for the availability, reliability, and simplicity of control [2,4]. Due to their intrinsic self-adaptability to the compliance of the biological tissues, non-conventional deformable actuators, such as shape memory alloy actuators [5], may be employed for this purpose. Pneumatic actuators provide some advantages over electric ones in terms of a high power-to-weight ratio, simplicity, safety, low cost, and easy maintenance. However, non-linearities due to air compressibility and actuator friction make it difficult to control pneumatic actuators [1]. Between the pneumatic actuators, the soft actuators present different advantages, due to very high force/weight ratio and low friction [6]. For all these reasons, among the pneumatic actuators, the McKibben pneumatic artificial muscle has been employed in several haptic devices, manipulators, and gloves [7–10].

The authors developed a prototype of a haptic finger, with a mechanism with one degree of freedom for force feedback actuated by McKibben's muscle. The device has been studied both as a rehabilitation system [11] and as a haptic system. In particular, a mathematical model of the device was developed, which was validated in a specific condition [12].

In the paper, the prototype of a haptic finger is presented, and the non-linear model developed is described. Therefore, further different validation tests of the model are presented. Finally, several simulations are presented, which are aimed at highlighting the dynamic behavior of the entire device.

2. The Prototype

Figure 1 shows the general scheme of the developed device. As regards the mechanism for the transmission of the reflection force on the fingertip, a one degree-of-freedom planar four-bar linkage has been chosen. The operator rests his hand on the fixed frame of the device, while his fingertip is held by a specially designed support integral with the beam force sensor mounted on the coupler of the four-bar mechanism (Figure 2). The operator, manipulating a virtual object or actuating a remote device, moves the finger between the position of maximum extension (Figure 2a) and that of maximum flexion (Figure 2b). For each position of the fingertip imposed by the operator, a corresponding angle α_1 of the rocker (Figure 1) is measured by a rotary encoder. The control unit, solving the forward kinematics equations of the mechanism, calculates the position of the fingertip (coordinates u, w in the sagittal plane, Figure 3a). Depending on the geometry and the stiffness of the remote or virtual object, the control unit evaluates the reflection force reference F_{ref} that must be applied to the fingertip. The control unit also calculates the error e between the force reference F_{ref} and the force F measured by the sensor, and provides the command signal for the pressure control proportional valve (FESTO® MPPE-3-1/4-10-010B) V_{ref}, which is used to regulate the upstream pressure p_1 of a fluidic resistance R. Regarding the actuation of the feedback force mechanism, a McKibben pneumatic muscle (Shadow Robot Company Ltd., London, UK) has been chosen. The pressure control proportional valve through the fluidic resistance R supplies the muscle and consequently generates a force F_{mu} that, in parallel with a spring force F_m, pulls a tendon that is connected to the coupler of the four-bar mechanism (Figures 1 and 2).

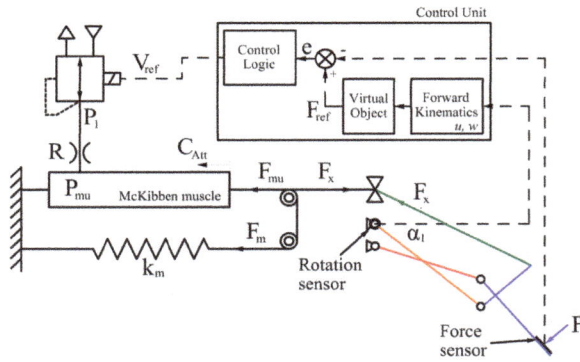

Figure 1. Scheme of the device.

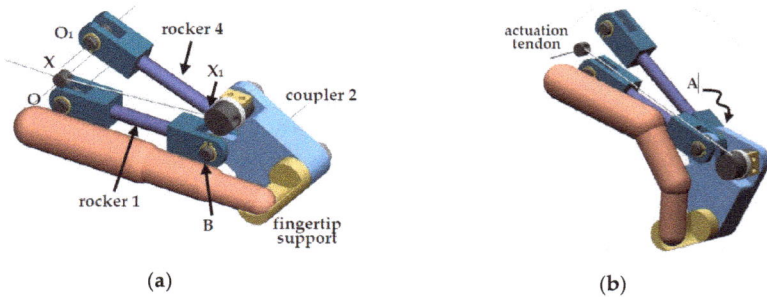

Figure 2. Mechanism for the transmission of the reflection force to the fingertip: (**a**) Finger at maximum extension; (**b**) finger at maximum flexion.

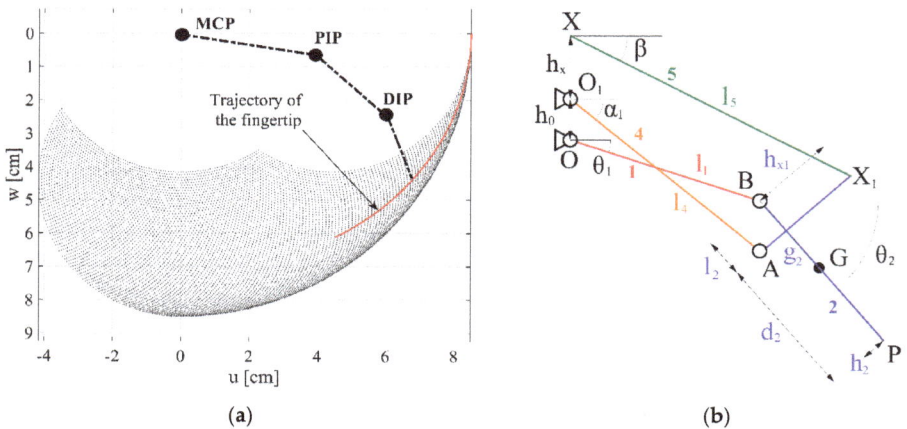

Figure 3. (**a**) Workspace of a human index fingertip. (**b**) Sketch of the mechanism (metacarpophalangeal joint, MCP, proximal interphalangeal joint, PIP, and distal interphalangeal joint, DIP).

The dimensional synthesis of the four-bar linkage was carried out by ensuring that the path generated by the coupler point P (fingertip holder) should be included into the natural workspace of a finger, which was calculated by taking into account the dependency among the rotations of each link of the finger (Figure 3a). The selected lengths of the proximal, middle, and distal phalanxes are respectively 4 cm, 2.5 cm, and 2 cm, while the rotation of the distal interphalangeal joint (DIP) joint has been considered to be equal to 2/3 of the proximal interphalangeal joint (PIP) joint rotation [11].

The dimensions of the links of the four-bar mechanism, whose scheme is shown in Figure 3b, are reported in Table 1. The operator's fingertip is positioned at point P, while the metacarpophalangeal joint (MCP) is coincident with the fixed hinge O. Tendon 5 is connected to coupler 2 in X_1; it passes through the low-friction support point X and is pulled by the McKibben muscle.

Table 1. Parameters of the model.

l_1	55 mm	h_{x1}	18 mm	n_t	1.52
l_2	5 mm	k_m	140 N/m	E	0.95 MPa
d_2	25 mm	F_{m0}	13.93 N	K_P	0.28–3 V/N
g_2	13 mm	C_{Att1}	20 mm	K_V	1×10^5 Pa/V
h_2	7 mm	m_c	55 g	ζ	1.8
l_4	63.9 mm	l_0	162 mm	σ_n	140 rad/s
l_{5max}	81.4 mm	l_i	110 mm	C	$5.25\ 10^{-9}$ m^3 Pa^{-1}s^{-1}
h_0	15 mm	r_i	10 mm	b	0.39
h_x	0 mm	s_i	0.8 mm	K_{obj}	8–50 N/rad

Figure 4 shows the prototype of the haptic finger that was realized and used for experimental tests. The labels refer to the Figures 2 and 3. The frame and the links of the mechanism were made of steel. Plain bearings were used to realize the revolute joints O, O_1, A, and B.

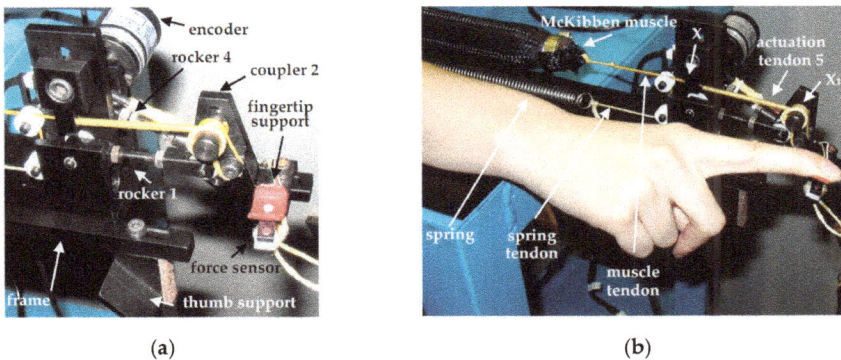

(a)   (b)

Figure 4. The prototype of the device. (**a**) Detail of the four-bar linkage and of the sensors. (**b**) Positioning of the operator's finger.

3. The Dynamic Model

The dynamic model of the haptic finger has been developed according to the logic represented in the block diagram of Figure 5.

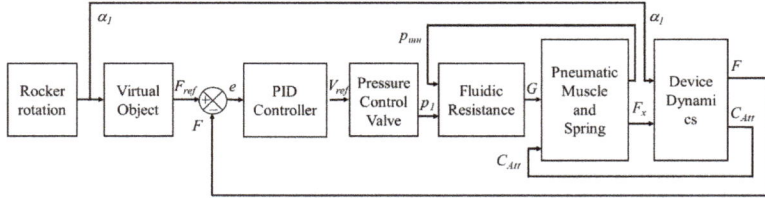

Figure 5. Block diagram of the dynamic model of the haptic finger.

Being known the rotation angle of the rocker α_1, depending on the position of the operator finger, and the force F_x generated by the McKibben muscle and by a parallel spring, the Device Dynamics block calculates the feedback force F that is applied to the fingertip and the stroke of the McKibben muscle C_{Att}.

The positioning in the sagittal plane (u, w) of each link of the mechanism can be calculated solving the non-linear system of Equations (1)–(3), thus obtaining ϑ_1, ϑ_2, and β:

$$l_4 \cos \alpha_1 + h_2 \sin \vartheta_2 - l_2 \cos \vartheta_2 = l_1 \cos \vartheta_1 \tag{1}$$

$$l_4 \sin \alpha_1 - h_2 \cos \vartheta_2 - l_2 \sin \vartheta_2 - h_0 = l_1 \sin \vartheta_1 \tag{2}$$

$$\beta = \tan^{-1} \left(\frac{h_x + h_0 + l_1 \sin \vartheta_1 + l_2 \sin \vartheta_2 - h_{x1} \cos \vartheta_2}{l_1 \cos \vartheta_1 + l_2 \cos \vartheta_2 + h_{x1} \sin \vartheta_2} \right) \tag{3}$$

Then, the trajectory of the fingertip in the (u, w) sagittal plane (Figure 3a) is calculated.

The stroke C_{Att} of the actuator depends on the length of the tendon l_5 and the preload stroke of the spring, C_{Att1}, and is calculated by solving Equation (4):

$$C_{Att} = l_{5max} - l_5 + C_{Att1}, l_5 = \frac{l_4 \cos \alpha_1 + (h_2 + h_{x1}) \sin \vartheta_2}{\cos \beta} \tag{4}$$

where l_{5max} corresponds to the maximum length XX_1 of the tendon (Figure 3b), which occurs at maximum finger flexion. The force F applied to the fingertip is calculated as a function of α_1, imposing the equilibrium of the coupler 2 (Figure 6). The direction of F is considered to be perpendicular to the coupler 2 in point P, corresponding to the fingertip support, due to low friction and consequent sliding between the fingertip and the same support. The equivalent mass of the system m_c has been considered as centered at point G belonging to the link 2. The static analysis yields the following Equations (5)–(7):

$$-F_B \cdot \cos \vartheta_1 + F_A \cdot \cos \alpha_1 - F_x \cdot \cos \beta = F \cdot \sin \vartheta_2 \tag{5}$$

$$F_B \cdot \sin \vartheta_1 - F_A \cdot \sin \alpha_1 + F_x \cdot \sin \beta = F \cdot \cos \vartheta_2 + m_c \cdot g \tag{6}$$

$$\begin{aligned}
F_A \cdot [\cos \alpha_1 \cdot (l_2 \cdot \sin \vartheta_2 + h_2 \cdot \cos \vartheta_2) - \sin \alpha_1 \cdot (l_2 \cdot \cos \vartheta_2 + h_2 \cdot \sin \vartheta_2)] \\
+F_x \cdot [\cos \beta \cdot (h_{x1} \cdot \cos \vartheta_2 - l_2 \cdot \sin \vartheta_2) + \sin \beta \cdot (h_{x1} \cdot \sin \vartheta_2 + l_2 \cdot \cos \vartheta_2)] \\
= F \cdot (l_2 + d_2) + m_c \cdot g \cdot \cos \vartheta_2 \cdot (g_2 + l_2)
\end{aligned} \tag{7}$$

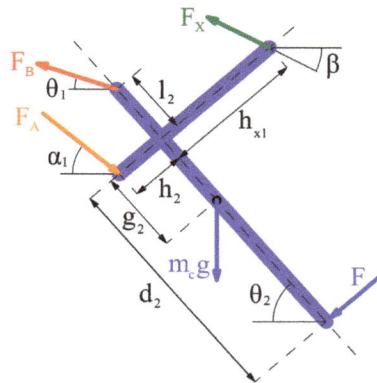

Figure 6. The free body diagram of the coupler (represented in the thick blue line).

Once the reflection force F has been calculated, and considering that: (i) the force reference F_{ref} is a function of the position (u, w) of the fingertip, depending on the geometry and the mechanical characteristic of the virtual/remote object, (ii) the fingertip position (u, w) is a function of the rocker rotation α_1, it is possible to evaluate the force error e:

$$e = F_{ref} - F \quad F_{ref} = F_{ref}(\alpha_1) \tag{8}$$

For the control, a proportional–integral–derivative (PID) algorithm has been adopted:

$$V_{ref} = K_P e + K_I \int_0^t e \, dt + K_D \frac{de}{dt} \tag{9}$$

where K_P, K_I, and K_D are, respectively, the proportional, integral, and derivative gains. After an iterative tuning approach, a simple proportional control came out to be sufficiently accurate and stable for the application, with a proportional coefficient of $K_P = 0.28$.

A second-order transfer function has been used to model the dynamic behavior of the pressure control proportional valve (*Pressure Control Valve* block):

$$p_1 = \frac{K_V \sigma_n^2}{s^2 + 2\zeta\sigma_n s + \sigma_n^2} V_{ref} \tag{10}$$

where K_V is the static gain, σ_n is the natural frequency, and ζ is the damping factor of the valve.

The pneumatic resistance (*Fluidic Resistance* block) has been modeled according to the ISO 6358 [13]. The mass air flow passing through the resistance can be calculated in sonic or subsonic condition, depending on the ratio between the downstream and upstream pressures:

$$G = \rho_0 P_1 C \; for \; 0 < \frac{P_{mu}}{P_1} \le b, \; G = \rho_0 P_1 C \sqrt{1 - \left(\frac{P_{mu}/P_1 - b}{1 - b}\right)^2} \; for \; b < \frac{P_{mu}}{P_1} \le 1 \tag{11}$$

where P_1 is the upstream absolute pressure, P_{mu} is the downstream absolute pressure, C is the sonic conductance, b is the critical ratio, and $\rho_0 = 1.18$ kg/m^3 is the air density in normal conditions. The C and b parameters are reported in Table 1.

Since $R = 287.2$ J/(kgK) is the air constant, and T is the absolute air temperature, assuming the variation of the internal volume V of the pneumatic muscle as negligible, the time derivative of the absolute internal pressure of the muscle for an isothermal transformation can be expressed as:

$$\frac{dP_{mu}}{dt} = \frac{RT}{V}G = \frac{RT}{\pi r^2 l}G \tag{12}$$

with the actual length l and the radius r of the actuator given by:

$$l = l_0 - C_{Att}, r = \frac{\sqrt{h^2 - l^2}}{2\pi n_t} \tag{13}$$

where, for the employed McKibben muscle (with a 20 mm diameter), C_{att} is the stroke of the pneumatic muscle, l_0 is the initial preloaded length, and n_t is the number of turns of the fibers, whose length h is 178 mm.

The internal relative pressure of the muscle p_{mu} can be calculated integrating Equation (12). In order to estimate the force exerted by the pneumatic actuator as a function of the internal pressure p_{mu}, a model of the McKibben muscle must be implemented. Various approaches have been proposed in the literature [14]. Ideal simple models consider only the kinematic relationship between the braided sheath and the inner tube, but unfortunately, they suffer when comparing theoretical and experimental results [15]. To overcome this issue, different models that account for the elastic energy that is stored in the bladder were developed based on the principle of virtual work [16–18]. Another approach to derive an analytical expression of the actuation force consists of writing balance equations of a free body diagram of the muscle [15]. Among these types of models, Ferraresi et al. [19] developed and validated a McKibben model, which is able to take into account the effects of the thickness and elasticity of the inner tube. Due to a good compromise between the simplicity and accurate numerical results, the latter model was used to calculate the force exerted by the muscle as a function of the internal pressure of the actuator [19]:

$$F_{mu} = -p_{mu}\frac{h^2 - l^2}{4\pi n_t^2} + E\frac{l - l_i}{l_i}2r_i\pi s_i + \left[p_{mu}l\frac{\sqrt{h^2 - l^2}}{2\pi n_t} - \frac{Es_i l}{r_i}\left(\frac{\sqrt{h^2 - l^2}}{2\pi n_t} - r_i\right)\right]\frac{l}{n_t\sqrt{h^2 - l^2}} \tag{14}$$

where r_i and l_i are the initial radius and length of the muscle at rest, respectively, s_i is the initial thickness of the inner chamber, and E is the Young modulus of the chamber.

Finally, the force F_x is directly connected to the force exerted by the muscle F_{mu} and the spring F_m by the following relations:

$$F_x = F_{mu} - F_m, F_m = k_m \cdot C_{Att} + F_{m0} \tag{15}$$

where k_m is the spring constant, and F_{m0} is its preload.

4. Experimental Validation of the Model

In order to verify the capability of the device to generate correct force feedback on the fingertip and validate the model, several experimental tests have been performed on the prototype presented in Section 2.

The angular rotation of the rocker 4 (Figure 3b) has been measured by a rotational encoder (BDK series, 1024 pulses, Baumer electric, Frauenfeld, Switzerland), which was processed and acquired by the incremental encoder interface board dSPACE® DS3002. The measurement of the force applied to the fingertip was made through a planar beam resistive sensor (Futek FR1020, capacity 89 N, combined error 0.25% rated output, Irvine, CA, USA) conditioned by a full bridge strain gauge module Meco, model MecoStrain. The acquisition of the force sensor signal was provided by a dSPACE® Multi-Channel A/D Board DS2002. The force sensor is mounted on the coupler by special grasping (Figure 7a). The fingertip support has been made by low-friction plastic material, as shown in Figure 7b, to ensure that the exchanged force can be considered perpendicular to the sensor itself.

The control algorithm was implemented in MATLAB-Simulink® environment, dSpace Control Desk. The command signal of the pressure control proportional valve V_{ref} was generated by a dSPACE® D/A board DS2101.

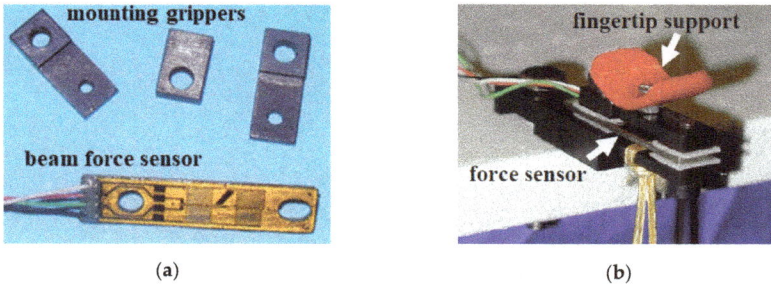

Figure 7. Force sensor. (**a**) The sensor and the mounting gripper. (**b**) The sensor in static calibration operation.

Figure 8a shows an image of the entire experimental setup during the static experimental tests.

In order to evaluate the dynamic performance of the haptic interface, several experimental tests were conducted. An operator, after positioning his hand on the device with a fully extended index finger, as shown in Figure 4b, is invited to freely perform flexions and extensions of the index finger (Figure 8b), as if he were virtually manipulating an object. The virtual object that was chosen for the experimental tests has a constant stiffness K_{obj}, so that the feedback force is proportional to the rotation of the rocker α_1 according to the equation $F_{ref} = K_{obj} \Delta\alpha_1$. During the manipulation, both the angular rotation of the rocker α_1, which is freely imposed by the operator, and the reflection force generated on the fingertip by the haptic device were acquired. In all of the tests, a purely proportional control ($K_P = 0.28$ V/N) was implemented, while the stiffness of the virtual object K_{obj} has been changed in the different tests, and fixed respectively equal to 8 N/rad, 10 N/rad, 20 N/rad, 30 N/rad, 40 N/rad, and 50 N/rad. Then, each test is characterized both by different temporal rotations of the rocker $\alpha_1 = \alpha_1(t)$ imposed by the operator, and by different stiffness K_{obj} assigned to the virtual object.

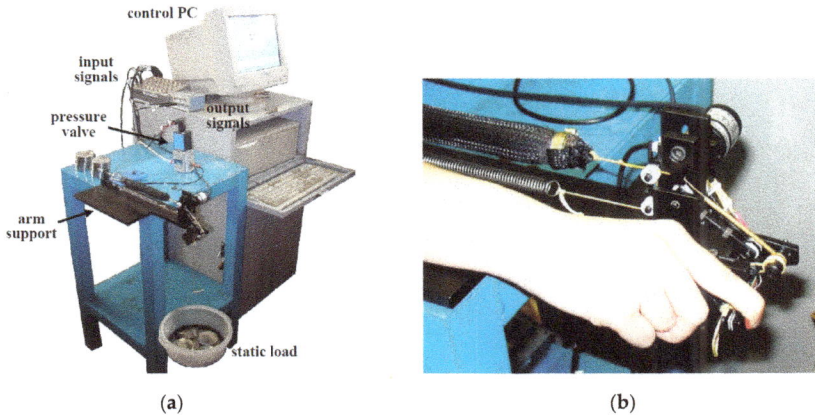

Figure 8. (**a**) Experimental setup. (**b**) Handling a virtual object.

In order to validate the model, simulations were performed by imposing, as an input, the same rotation law of rocker $\alpha_1 = \alpha_1(t)$ recorded in each experimental test, and calculating the reflection force generated by the system with the same control (purely proportional control with $K_P = 0.28$ V/N), and with same stiffness of the virtual object K_{obj}.

Figure 9 shows the experimental results obtained in the different tests described above, compared with the results of the simulation conducted under the same conditions. The figures show the trends of the experimental and simulated feedback force as a function of time, for different stiffnesses of the object, between 8 N/rad and 50 N/rad.

The most relevant differences between the simulated and the experimental results occurred for more rigid virtual objects, in cases where the operator imposed low-rocker rotation values (low reaction forces) and quickly reversed the direction of rotation.

Nevertheless, the model highlights the ability to predict with good accuracy the trend of the force generated by the device in dynamic conditions.

Based on these results, the model was considered validated and sufficiently reliable to predict the dynamic performance of the device, even under conditions different from those tested. This analysis is presented and discussed in the next section.

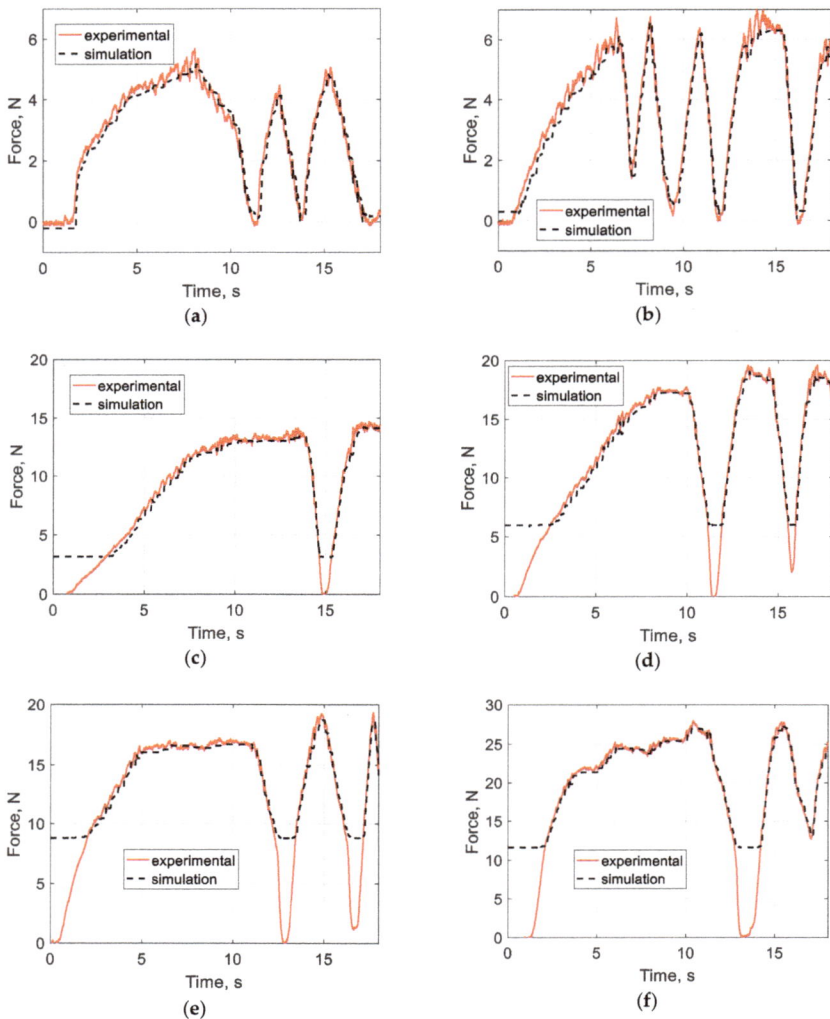

Figure 9. Comparison between experimental and simulation results. (a) $K_{obj} = 8$ N/rad. (b) $K_{obj} = 10$ N/rad. (c) $K_{obj} = 20$ N/rad. (d) $K_{obj} = 30$ N/rad. (e) $K_{obj} = 40$ N/rad. (f) $K_{obj} = 50$ N/rad.

5. Dynamic Assessment of the Device

The validated model was used to evaluate the dynamic performance of the device. A first series of simulations concerned the study of the response to step input, for different proportional gain K_P values of the pure proportional control (Figure 10). Due to pure proportional control, a steady state error is observed. As expected, when increasing the proportional gain K_P, the steady-state error decreases, but for proportional gain values close to 2 V/N, the system tends to instability, and in fact becomes unstable for K_P values equal to 3 V/N.

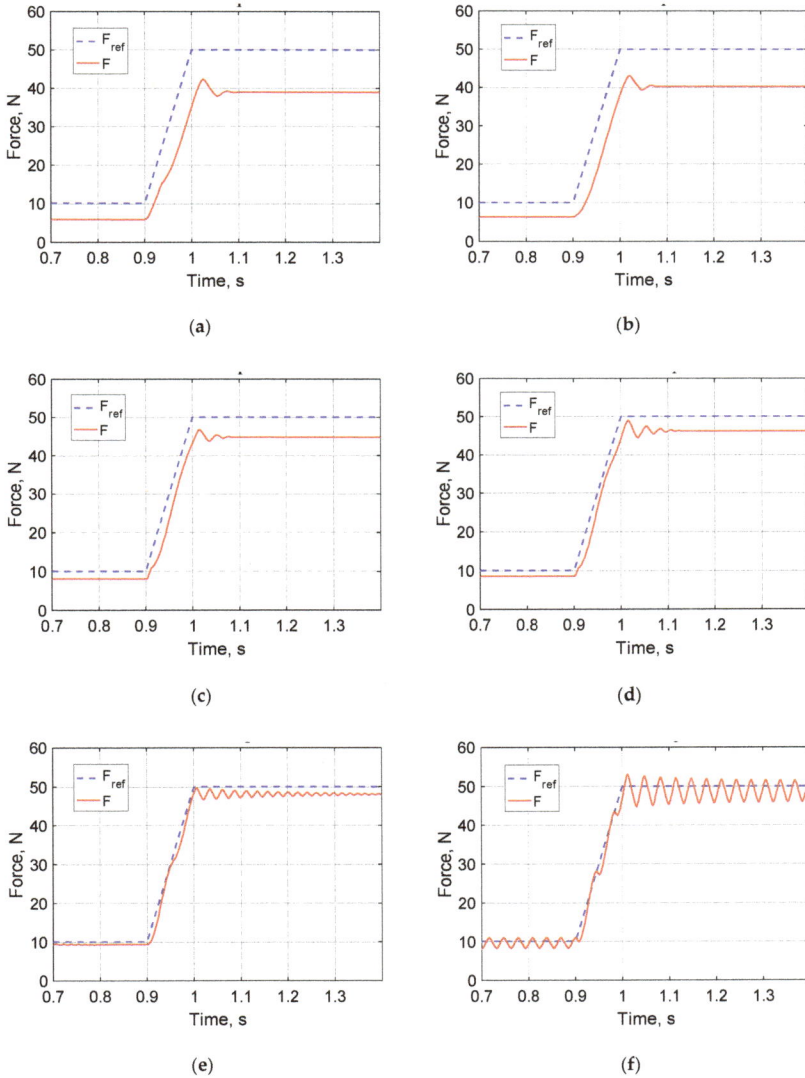

Figure 10. Model step response. (a) K_P = 0.3 V/N. (b) K_P = 0.35 V/N. (c) K_P = 0.7 V/N. (d) K_P = 1 V/N. (e) K_P = 2 V/N. (f) K_P= 3 V/N.

The bandwidth of the device was also estimated using the dynamic model, again with a purely proportional control. Figure 11 shows the Bode diagram of the system, whose behavior has been simulated under nominal conditions ($K_P = 0.7$ V/N). We estimated a bandwidth close to 80 rad/s, which is certainly compatible with many haptic applications.

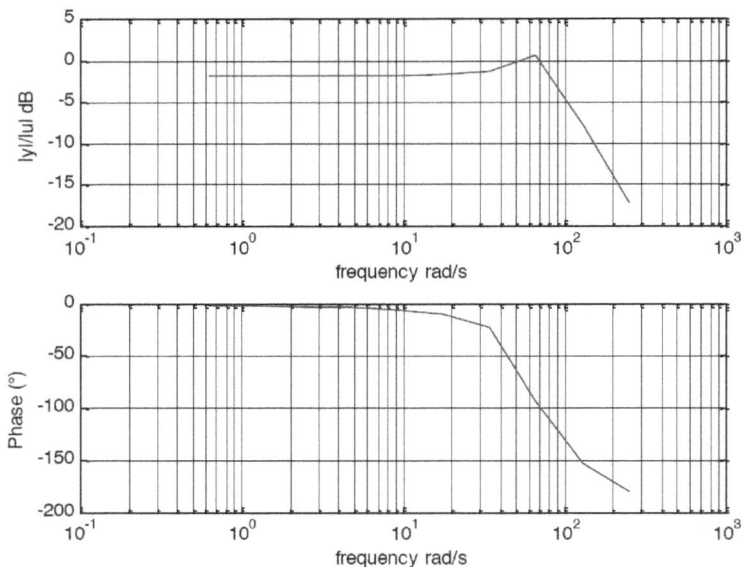

Figure 11. Model frequency response in terms of F/F_{ref} ($K_P = 0.7$ V/N).

6. Conclusions

A novel haptic device, which was conceived as a finger for a haptic glove, has been developed and tested. The device is based on a four-bar mechanism, and is actuated by a McKibben muscle. The work has been particularly focused on the employment of such a kind of actuator in haptic applications.

A prototype has been realized and analytically modeled. The model, which was experimentally validated, proved to be able to predict with good accuracy the behavior of the system, and was used both to define the control algorithm and assess the static and dynamic performance of the device.

Experimental tests on the prototype highlighted its effectiveness in applications requiring good haptic sensitivity, confirming that the McKibben muscle can provide relevant advantages when used in these kinds of applications.

Future work will be focused first on the improvement of the control, in particular introducing an integral gain in order to achieve zero steady-state error, and then on the optimization of the finger's mechanical structure, with the aim of realizing a full haptic glove.

Author Contributions: Conceptualization, W.F. and C.F.; methodology, D.M. and C.D.B.; software, W.F. and C.D.B.; validation, C.F. and D.M.; writing—original draft preparation, W.F.; writing—review and editing, C.D.B.; visualization, D.M.; supervision, C.F.

Funding: This research has not received external funding.

Conflicts of Interest: The authors declare no conflict of interest.

References

1. Uddin, M.W.; Zhang, X.; Wang, D. A pneumatic-driven haptic glove with force and tactile feedback. In Proceedings of the 2016 International Conference on Virtual Reality and Visualization (ICVRV2016), Hangzhou, China, 24–26 September 2016; pp. 304–311.
2. Heo, P.; Gu, G.M.; Lee, S.; Rhee, K.; Kim, J. Current hand exoskeleton technologies for rehabilitation and assistive engineering. *Int. J. Precis. Eng. Manuf.* **2012**, *13*, 807–824. [CrossRef]
3. Malvezzi, M.; Valigi, M.C.; Salvietti, G.; Iqbal, Z.; Hussain, I.; Prattichizzo, D. Design Criteria for Wearable Robotic Extra–Fingers with Underactuated Modular Structure. In Proceedings of the 2nd IFToMM ITALY Conference, Cassino, Italy, 29–30 November 2018; pp. 509–517.
4. Gerding, E.C.; Carbone, G.; Cafolla, D.; Russo, M.; Ceccarelli, M.; Rink, S.; Corves, B. Design and Testing of a Finger Exoskeleton Prototype. In Proceedings of the 2nd IFToMM ITALY Conference, Cassino, Italy, 29–30 November 2018; pp. 342–349.
5. Maffiodo, D.; Raparelli, T. Three-Fingered Gripper with Flexure Hinges Actuated by Shape Memory Alloy Wires. *Int. J. Autom. Technol.* **2017**, *11*, 355–360. [CrossRef]
6. Ferraresi, C.; Franco, W.; Quaglia, G. A novel bi-directional deformable fluid actuator. *J. Mech. Eng. Sci.* **2014**, *228*, 2799–2809. [CrossRef]
7. Sun, Z.; Bao, G.; Yang, Q.; Wang, Z. Design of a Novel Force Feedback Dataglove Based on Pneumatic Artificial Muscles. In Proceedings of the International Conference on Mechatronics and Automation, Luoyang, China, 25–28 June 2006; pp. 968–972.
8. Li, H.; Kawashima, K.; Todano, K.; Ganguly, S.; Nakano, S. Achieving Haptic Perception in Forceps' Manipulator Using Pneumatic Artificial Muscle. *IEEE/ASME Trans. Mechatron.* **2013**, *18*, 74–85. [CrossRef]
9. Moon, K.; Ryu, D.; Chun, C.; Lee, Y.; Kang, S.; Park, M. Development of a Slim Haptic Glove Using McKibben Artificial Muscles. In Proceedings of the SICE-ICASE International Joint Conference, Bexco, Busan, Korea, 18–21 October 2006; pp. 204–208.
10. Egawa, M.; Watanabe, T.; Nakamura, T. Development of a Wearable Haptic Device with Pneumatic Artificial Muscles and MR brake. In Proceedings of the IEEE Virtual Reality (VR), Arles, France, 23–27 March 2015; pp. 173–174.
11. De Benedictis, C.; Ferraresi, C.; Franco, W.; Maffiodo, D. Hand rehabilitation device actuated by a pneumatic muscle. In Proceedings of the 27th International Conference on Robotics in Alpe-Adria-Danube Region (RAAD 2018), Patras, Greece, 6–8 June 2018; pp. 102–111.
12. Franco, W.; Maffiodo, D.; De Benedictis, C.; Ferraresi, C. Dynamic modeling and experimental validation of a haptic finger based on a McKibben muscle. In Proceedings of the 4th IFToMM Symposium on Mechanism Design for Robotics, Udine, Italy, 11–13 September 2018; pp. 251–259.
13. Ferraresi, C.; Sorli, G. Modelling of pneumatic systems. *Fluid Apparecchiature Idraul. Pneum.* **1993**, *346*, 48–54.
14. Tondu, B. Modelling of McKibben artificial muscle: A review. *J. Intell. Mater. Syst. Struct.* **2012**, *23*, 225–253. [CrossRef]
15. Kothera, C.S.; Jangid, M.; Sirohi, J.; Wereley, N.M. Experimental characterization and static modeling of McKibben Actuators. *J. Mech. Des.* **2009**, *131*, 091010. [CrossRef]
16. Klute, G.; Hannaford, B. Accounting for Elastic Energy Storage in McKibben Artificial Muscle Actuators. *ASME J. Dyn. Syst. Meas. Control* **2000**, *122*, 386–388. [CrossRef]
17. Tsagarakis, N.; Caldwell, D. Improved Modelling and Assessment of Pneumatic Muscle Actuators. In Proceedings of the IEEE International Conference on Robotics and Automation (ICRA 2000), San Francisco, CA, USA, 24–28 April 2000; pp. 3641–3646.
18. Sugimoto, Y.; Naniwa, K.; Osuka, K.; Sankai, Y. Static and dynamic properties of McKibben pneumatic actuator for self-stability of legged robot motion. *Adv. Robot.* **2013**, *27*, 469–480. [CrossRef]
19. Ferraresi, C.; Franco, W.; Manuello Bertetto, A. Flexible Pneumatic Actuators: A comparison between the McKibben and the straight fibres muscle. *J. Robot. Mechatron.* **2001**, *13*, 56–63. [CrossRef]

![robotics logo] *robotics*

MDPI

Article

Singularity Avoidance Control of a Non-Holonomic Mobile Manipulator for Intuitive Hand Guidance [†]

Matthias Weyrer [1],*, Mathias Brandstötter [1] and Manfred Husty [2]

[1] Joanneum Research, Institute for Robotics and Mechatronics, Lakeside B08a, 9020 Klagenfurt am Wörthersee, Austria; mathias.brandstoetter@joanneum.at

[2] Unit Geometry and CAD, University of Innsbruck, 6020 Innsbruck, Austria; manfred.husty@uibk.ac.at

* Correspondence: matthias.weyrer@joanneum.at; Tel.: +43-316-876-2010

† This paper is an extended version of our paper published in Weyrer, M.; Brandstötter, M.; Mirkovic, D. Intuitive Hand Guidance of a Force-Controlled Sensitive Mobile Manipulator. In Proceedings of the IFToMM Symposium on Mechanism Design for Robotics, Udine, Italy, 11–13 Augus 2018.

Received: 20 January 2019; Accepted: 14 February 2019; Published: 19 February 2019

Abstract: Mobile manipulators are robot systems capable of combining logistics and manipulation tasks. They thus fulfill an important prerequisite for the integration into flexible manufacturing systems. Another essential feature required for modern production facilities is a user-friendly and intuitive human-machine interaction. In this work the goal of code-less programming is addressed and an intuitive and safe approach to physically interact with such robot systems is derived. We present a natural approach for hand guiding a sensitive mobile manipulator in task space using a force torque sensor that is mount close to the end effector. The proposed control structure is capable of handling the kinematic redundancies of the system and avoid singular arm configurations by means of haptic feedback to the user. A detailed analysis of all possible singularities of the UR robot family is given and the functionality of the controller design is shown with laboratory experiments on our mobile manipulator.

Keywords: robot kinematics; robot singularity; singularity analysis; robot control; mobile manipulation; human-robot-interaction; learning by demonstration; compliance control

---

1. Introduction

The demand for highly flexible and adaptable robotic systems naturally arises within the manufacturing processes of products with high variability and small lot-sizes. This challenges also include frequently reprogramming of the robot. Traditionally, interactions between humans and robots within a shared workplace can be categorized into two distinct scenarios: a service scenario and a process scenario. In the former case, a robot is programmed and prepared for a new production process rather infrequently by highly skilled experts. In the latter case, less complicated interactions are part of the everyday work flow. This means that robot reprogramming has to be performed much more frequently by human workers with extensive domain knowledge but usually limited programming skills.

To integrate this reprogramming fluently into the workflow it must be fast and easy to use. Thus the interaction interface between human and robot is of significant importance. One well known technique is Programming by Demonstration (PbD). There are several forms of this method: (a) the positions of the work-piece itself or a special teaching object is tracked and used to plan the trajectory of the robot [1], (b) the robot is guided into the desired positions via remote control [2,3] and (c) kinestetic programming by demonstration, where the robot is compliant and can be hand-guided into the desired configurations [4,5]. In the user-centric work of [6], the trajectories that are taught to the robot system by untrained end users can be adapted in a subsequent step via a graphical user interface to obtain the desired task.

The latter mentioned teaching technique requires the robot to be compliant. There are several different sensitive robots, also known as collaborative robots or cobots that are able to perceive the interaction forces with the environment by utilizing additional sensing like joint torque sensors and control theory. With the knowledge of external forces that act on the robot, a compliant behavior can be realized which enables the ability to hand-guide the robot and allows a closer cooperation without external safety barriers [7].

While there are many publications describing compliance control for serial manipulators, e.g., [8–13] only little investigations for a whole-body compliance control of a mobile manipulators have been done. Leboutet et al. [14] proposed a technique with hierarchical force propagation for a mobile manipulator that consists of an omni-directional base and two Universal Robots UR10 serial robots. The robotic arms are covered with their special multi-modal sensor skin which allows measuring the applied external forces on the robot at several contact points. External forces whose reactive motions are inconvenient to be performed by the serial manipulator are directly projected to the mobile base. To decide which motions should be performed by the base, the manipulability ellipsoid is used. Navarro et al. [15] presented a solution for an omnidirectional base where the distribution of motion is done with optimization. They proposed a cost function that includes a measure for the manipulability, a self-defined value for the closeness-to-singularity and some additional distance and angle constraints.

Han et al. [16] point out the complexity of controlling a robot in task-space while taking singularities and joint limits into account. They present a hierarchically structured controller that uses a continuous task transition algorithm to guarantee execution of the main task while additional tasks, e.g., for singularity-avoidance, can be activated or deactivated.

In our previous work [17], we presented a control design for a whole body compliance control of the mobile manipulator but singularity avoidance was not taken into account. Since we control the velocities of the end effector (EE) in task space, singular configurations are problematic. In a singular configuration the inverse kinematic on velocity level cannot be solved at all or results in infinity joint velocities. Also approaching a configuration close to a singularity may result in very high joint speeds, which could be dangerous for humans close to the robot, and must be avoided. We extended our previous work by analyzing all possible singularities of the Universal Robots family with focus on the model UR10, which is used on our mobile manipulator CHIMERA. We also included a singularity-avoidance strategy in our control structure by applying haptic feedback to the user before approaching singular configurations and present the results of conducted laboratory experiments.

This paper is organized as follows: The kinematics and especially the singularity analysis of the serial manipulator UR10 is given in Section 2, the control structure is discussed in Section 3, including the motion-distribution between mobile base and serial manipulator and our proposed strategy to avoid approaching singular arm configurations. Experimental results are shown in Section 4 and a conclusion and outlook for future work is given in Section 5.

2. Kinematics and Singularity Analysis

A mobile manipulator is an effective tool to accomplish tasks, e.g. the manipulation of objects in space. It is a combination of a serial manipulator and a mobile robot, which greatly expands the manipulator's workspace and thus increases the system's performance. For analysis purposes, such systems can often be split into two components, a mobile platform and a manipulator arm. The studies in this paper focus on a mobile manipulator called CHIMERA, which consists of a MiR platform (differential drive) and a UR10 (6 DoF) serial arm.

2.1. Kinematics

Mobile wheeled platforms have been the subject of many studies in the past. For the kinematic description of mobile robots we refer to [18]. The kinematic relationships of the UR10 were also sufficiently investigated [19], although it is pointed out that the kinematic chain has an offset wrist.

2.2. Singularity Analysis of the UR Robot

For the computation of all singularities of the UR10 we will use the well known fact that the columns of the 6×6 Jacobian matrix $\mathbf{J}$ are the Plücker coordinates of the instantaneous locations of the rotation axes of the manipulator [20]. Using this fact one can obtain $\mathbf{J}$ without differentiation. A couple of prerequisites are noted before. We assume that the rotation axes are always the z-axes of the local coordinate systems. In this local coordinate system the Plücker coordinates of the revolute axes are $\mathbf{p}_i = [0, 0, 1, 0, 0, 0]^T$. To compute their coordinates in the base system the forward transformation matrices are needed. It has to be noted that the manipulator is in a singular pose when the six Plücker coordinates are linearly dependent.

Using the usual Denavit-Hartenberg (D-H) convention to describe the geometric structure of the serial manipulator [21], the forward transformation can be written as

$$\mathbf{T} = \prod_{i=1}^{6} \mathbf{M}_i \cdot \mathbf{G}_i \tag{1}$$

where

$$\mathbf{M_i} = \begin{bmatrix} 1 & 0 & 0 & 0 \\ 0 & \cos q_i & -\sin q_i & 0 \\ 0 & \sin q_i & \cos q_i & 0 \\ 0 & 0 & 0 & 1 \end{bmatrix}, \ \mathbf{G_i} = \begin{bmatrix} 1 & 0 & 0 & 0 \\ a_i & 1 & 0 & 0 \\ 0 & 0 & \cos \alpha_i & -\sin \alpha_i \\ d_i & 0 & \sin \alpha_i & \cos \alpha_i \end{bmatrix}.$$

The joint positions of the serial manipulator are given by q_i as depicted in Figure 2 and the constant D-H parameters are given by a_i, d_i and α_i. To transform the Plücker coordinates the line transform matrix $\overline{\mathbf{T}}$ is needed. When the forward transformation matrix is written as

$$\mathbf{T} = \begin{bmatrix} 1 & 0 \\ \mathbf{a} & \mathbf{A} \end{bmatrix}, \quad \mathbf{a} \dots 3 \times 1 \text{ translation vector, } \mathbf{A} \dots 3 \times 3 \text{ rotation matrix}$$

then the line transform matrix is

$$\overline{\mathbf{T}} = \begin{bmatrix} \mathbf{A} & 0 \\ \mathbf{a}^\times \mathbf{A} & \mathbf{A} \end{bmatrix}. \quad \mathbf{a}^\times \dots \text{skew symmetric matrix belonging to translation vector } \mathbf{a}$$

To compute the Plücker coordinates of a specific rotation axis only those parts of the forward kinematics will be needed which transform up the axis whose location has to be found. We denote the partial transformations by

$$\mathbf{T}_j = \prod_{i=1}^{j} \mathbf{M}_i \cdot \mathbf{G}_i, \quad j = 1, \dots, 5$$

and by $\mathbf{y}_1 = [0, 0, 1, 0, 0, 0]^T$ the Plücker coordinates of the first rotation axis. Then the remaining five Plücker coordinates are obtained by

$$\mathbf{y}_k = \overline{\mathbf{T}}_{k-1} \cdot \mathbf{y}_1. \quad k = 2, \dots, 6 \tag{2}$$

The six Plücker coordinates can now be assembled to the 6×6 Jacobian matrix $\mathbf{J}$:

$$\mathbf{J} = [\mathbf{y}_1, \mathbf{y}_2, \mathbf{y}_3, \mathbf{y}_4, \mathbf{y}_5, \mathbf{y}_6] \tag{3}$$

A necessary and sufficient condition for the manipulator being in a singularity is: $\det \mathbf{J} = 0$. Due to the simplicity of the design of the manipulator this determinant can be computed without assigning all D-H parameters. The resulting equation becomes very well laid out when all angles in the forward transformation are written in algebraic values. This is achieved by performing half tangent substitution: $\cos q_i = \frac{1-v_i^2}{1+v_i^2}$, $\sin q_i = \frac{2v_i}{1+v_i^2}$, $\cos \alpha_i = \frac{1-al_i^2}{1+al_i^2}$, $\sin \alpha_i = \frac{2al_i}{1+al_i^2}$. The essential D-H parameters that determine the UR family of robots are $a_1 = 0, d_2 = 0, d_3 = 0, a_4 = 0, a_5 = 0,$ $a_6 = 0, al_1 = 1, al_2 = 0, al_3 = 0, al_4 = -1, al_5 = 1, al_6 = 0$. The remaining D-H parameters are not assigned and determine the type of UR robot. Computing the determinant of $\mathbf{J}$ yields

$$
\begin{aligned}
\det \mathbf{J} = v_3 v_5 \Big[& (v_4^2 + 1)(v_3^2 + 1)(v_2 - 1)(v_2 + 1)a_2 - (v_4^2 + 1)(v_2 v_3 + v_2 + v_3 - 1)(v_2 v_3 - v_2 - v_3 - 1)a_3 \\
& - (2(v_2 v_3 + v_2 v_4 + v_3 v_4 - 1))(v_2 v_3 v_4 - v_2 - v_3 - v_4)d_5 \Big] = 0.
\end{aligned}
\tag{4}
$$

The analysis of Equation (4) reveals that $\det \mathbf{J}$ factors into three parts: $v_3 = 0$ determines the elbow singularities because then the arm is stretched out, $v_5 = 0$ yields the wrist singularities because then the fourth and the sixth axis are coplanar. The third expression belongs to the shoulder singularity and contains only the joint parameters v_2, v_3, v_4. When two of the three joint parameters are set, then the third can be computed via the remaining quadratic equation. When the manipulator is brought to the resulting pose then one can observe that the intersection point P_{56} of the fifth and the sixth axis is on a cylinder which has the equation $x^2 + y^2 - d_4^2 = 0$ in the base coordinate system. This cylinder has a geometrical easy explanation: lets assume for a moment $v_1 = 0$, then it is obvious that the intersection point of fifth and sixth axis can only move in the plane $y = -d_4$ of the base coordinate system. This plane intersects the plane $x = 0$ which is the span of the first and the second axis in a line parallel to the z-axis in a distance d_4 from this axis. When the rotation about the first axis is added then this line describes the cylinder. That P_{56} is located on this line in case of a shoulder singularity can be computed immediately by setting $v_1 = 0$ and solving the third polynomial of Equation (4) for, e.g., $v_4 = f(v_2, v_3)$. As the equation is quadratic in v_4 one obtains for arbitrary values of v_3 and v_4 two values for $v_4 = v_{41}, v_4 = v_{42}$. Direct computation of the location of P_{56} when either v_{41} or v_{42} are substituted into the forward kinematic equation yields $P_{56} = [1, 0, -d_4, \pm g(v_2, v_3)]^T$. This shows that P_{56} is on the intersection line of planes $x = 0$ and $y = -d_4$. Its z coordinate is determined by $g(v_2, v_3)$ which is a relatively complicated function. It gives the values of the intersection point of the circle which is the path of P_{56} during the rotation about the fourth axis with the plane $x = 0$.

The forgoing description is valid for all manipulators of the UR family. When a special type is chosen, e.g. UR10, then the remaining D-H parameters are set $a_2 = 0.6127$, $a_3 = 0.5716$, $d_1 = 0.118$, $d_4 = 0.163941$, $d_5 = 0.1157$, $d_6 = 0.0922$ and the singularity equation becomes:

$$
\begin{aligned}
\det \mathbf{J} = v_3 v_5 \Big[& 0.6127(v_4^2 + 1)(v_3^2 + 1)(v_2 - 1)(v_2 + 1) - \Big] \\
& 0.5716(v_4^2 + 1)(v_2 v_3 + v_2 + v_3 - 1)(v_2 v_3 - v_2 - v_3 - 1) \\
& - 0.2314(v_2 v_3 + v_2 v_4 + v_3 v_4 - 1)(v_2 v_3 v_4 - v_2 - v_3 - v_4) \Big] = 0
\end{aligned}
\tag{5}
$$

The singularity surface represented by Equation (5) is shown in Figure 1.

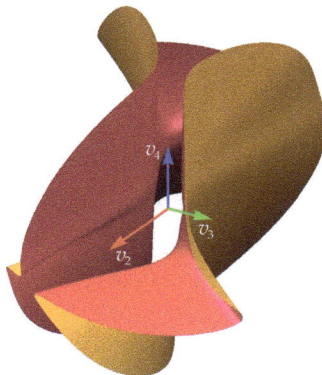

Figure 1. Singularity surface of shoulder singularities in the transformed joint space.

3. Control Strategy

The goal of the control structure is to translate the input forces and torques of the user into robot motion. We divide between the two subsystems, the mobile base and the serial manipulator on top of it. Since the combined system shows kinematic redundancies concerning the 3D task-space, the motion distribution is a main part of the proposed control structure. Additionally, virtual springs are used to generate haptic feedback to the user when pushing or pulling the mobile base. Haptic feedback is also used to avoid singular arm configurations.

We consider the serial manipulator as an open kinematic chain with $\mathbf{q}_{\text{ur}} = \begin{bmatrix} q_1 & q_2 & \cdots & q_6 \end{bmatrix}^T \in \mathbb{R}^{6 \times 1}$ joints on top of the mobile base equipped with a differential drive, denoted as $\mathbf{q}_{\text{mir}} = \begin{bmatrix} x & y & \theta \end{bmatrix}^T \in \mathbb{R}^{3 \times 1}$ shown in Figure 2. All freedoms of the system are collected in $\mathbf{q}_{\text{sys}} = \begin{bmatrix} \mathbf{q}_{\text{ur}}^T & \mathbf{q}_{\text{mir}}^T \end{bmatrix}^T \in \mathbb{R}^{9 \times 1}$. Moreover, the redundant robot system is considered as a unit that is composed of two tightly coupled subsystems, where the coupling is established by our proposed control structure.

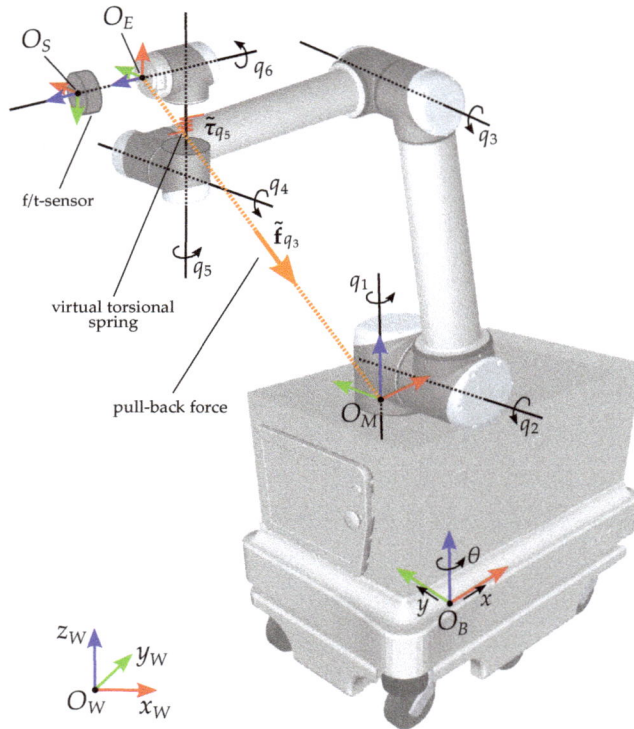

Figure 2. CHIMERA joints and coordinates: The mobile base is modelled with two linear joints x and y and one rotational joint Θ. The UR-10 has six rotational joints denoted as q_i with $i = 1, \ldots, 6$. The Coordinate Systems are defined with their origins O and three axis-vectors x, y and z. Shown are the world-coordinate system $\Sigma_W := \{O_W; x_W, y_W, z_W\}$, the frame of the mobile base $\Sigma_B := \{O_B; x_B, y_B, z_B\}$, the UR-10 base frame $\Sigma_M := \{O_M; x_M, y_M, z_M\}$, the EE frame $\Sigma_E := \{O_E; x_E, y_E, z_E\}$ and the coordinate system of the force-torque sensor $\Sigma_S := \{O_S; x_S, y_S, z_S\}$. The virtual pull-back force for singularity avoidance in joint 3 is denoted as $\tilde{\mathbf{f}}_{q_3}$ and the virtual torque for singularity avoidance in joint 5 as τ_5.

3.1. Distribution of Motion

The distribution of motion is realized as follows: Two circles, an inner and an outer one, are used to define three zones in the xy-plane of the robot base frame, as depicted in Figure 3. We switch between three main operation modes, depending on the position of the end effector (EE) in the xy-plane. If the EE is located between the two circles ($r_i < r < r_o$), only the serial manipulator moves, denoted as *UR-Mode*. Outside of the outer circle ($r > r_o$) we switch to *Pull-Mode*, where the mobile base can be pulled like a trailer and haptic user-feedback is realized by means of a virtual spring. This virtual spring generates a force to move the EE back inside the circle. When the EE enters the inner circle ($r < r_i$)we switch to *Push-Mode*. The user can move the base by pushing it and a virtual spring generates a force to move the EE back out of this inner circle.

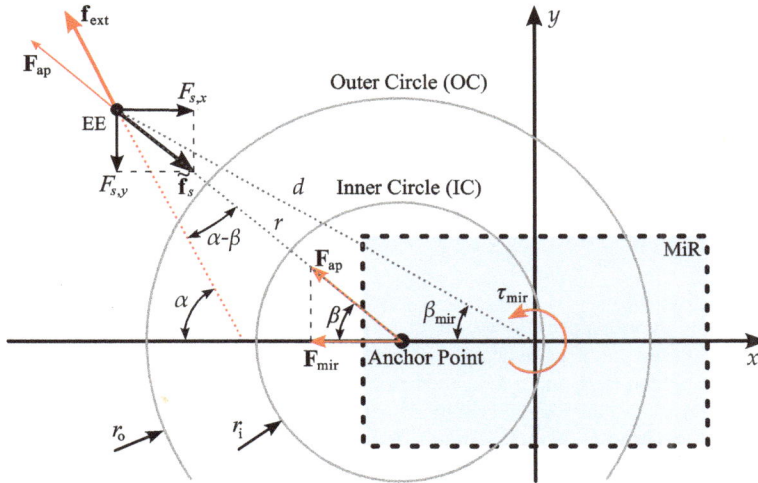

Figure 3. Kinetic relationships under external forces: This figure illustrates the angles and forces when the EE is outside the outer circle (*Pull-Mode*) and a force and torque is projected to the mobile base as described in Section 3.1.

The control inputs of the system are the EE-velocities $\dot{\mathbf{x}}_{ur}^{\Sigma_B} = \begin{bmatrix} \mathbf{v}^T & \boldsymbol{\omega}^T \end{bmatrix}^T \in \mathbb{R}^{6 \times 1}$ and the velocities of a mobile base in the general case $\dot{\mathbf{q}}_{mir}^{\Sigma_B} = \begin{bmatrix} \dot{x} & \dot{y} & \dot{\Theta} \end{bmatrix}^T \in \mathbb{R}^{3 \times 1}$, all given in the frame of the mobile base $\Sigma_B := \{ O_B; x_B, y_B, z_B \}$. For simplicity, we drop the subscript for the reference coordinate, thus in the following, vectors without an explicit subscript are all given in the mobile-base-frame. In all modes, the controller equations are given by

$$\begin{bmatrix} \dot{\mathbf{x}}_{ur} \\ \dot{\mathbf{q}}_{mir} \end{bmatrix} = \begin{bmatrix} \mathbf{B}_{ur}^{-1} & \mathbf{0} \\ \mathbf{0} & \mathbf{B}_{mir}^{-1} \end{bmatrix} \begin{bmatrix} \mathbf{w}_{ext} + \tilde{\mathbf{w}}_{fb} \\ \tilde{\mathbf{w}}_{mir} \end{bmatrix} - \begin{bmatrix} \dot{\mathbf{x}}_c \\ \mathbf{0} \end{bmatrix} \tag{6}$$

where $\mathbf{B}_{ur} \in \mathbb{R}^{6 \times 6}$ and $\mathbf{B}_{mir} \in \mathbb{R}^{3 \times 3}$ are the diagonal positive definite damping matrices, $\mathbf{w}_{ext} = \begin{bmatrix} \mathbf{f}_{ext}^T & \boldsymbol{\tau}_{ext}^T \end{bmatrix}^T \in \mathbb{R}^{6 \times 1}$ is the wrench vector, including external forces and torques applied to the EE, $\tilde{\mathbf{w}}_{fb} \in \mathbb{R}^{6 \times 1}$ is a wrench vectors for haptic feedback including the virtual spring forces and singularity avoidance wrenches as described in Section 3.2 and $\tilde{\mathbf{w}}_{mir} = \begin{bmatrix} F_{mir} & 0 & \tau_{mir} \end{bmatrix}^T \in \mathbb{R}^{3 \times 1}$ includes the projected force for linear motion and projected torque for angular motion of the mobile base as shown in Equation (7). The vector of EE-velocities to compensate for angular motions of the base is denoted as $\dot{\mathbf{x}}_c$. We assume that the applied wrench $\mathbf{w}_{ext}$ acting on the EE is known, either by using a force-torque sensor or joint torque estimation based on motor current measurements (see, e.g., [22,23]).

Mode-dependent variables are the projected wrench $\tilde{\mathbf{w}}_{\text{mir}}$ of the mobile base and the haptic feedback wrench $\tilde{\mathbf{w}}_{\text{fb}}$. To move the mobile base, we project the applied external wrench to a linear pulling or pushing force F_{mir} and a rotation torque τ_{mir}. These projected values are only computed if the EE is not located in between the inner and the outer circle, e.g., in *Pull-Mode* and *Push-Mode*. Since the mobile base is non-holonomic due its the differential drive, no linear motion in y-direction is possible and the second entry of the projected wrench $\tilde{\mathbf{w}}_{\text{mir}} = \begin{bmatrix} F_{\text{mir}} & 0 & \tau_{\text{mir}} \end{bmatrix}^{\text{T}} \in \mathbb{R}^{3\times1}$ is set to zero. This strategy is inspired by the design of a steered trailer, which most persons are familiar with. The projections are given as

$$
\begin{bmatrix} F_{\text{mir}} \\ \tau_{\text{mir}} \end{bmatrix} = \begin{cases} \begin{bmatrix} |\mathbf{f}_{\text{ext}}|\cos(\alpha-\beta)\sin(\beta) \\ p_x|\mathbf{f}_{\text{ext}}|\cos(\alpha-\beta)\cos(\beta) \end{bmatrix} & r > r_{\text{out}} \text{ and } |\beta|-|\alpha| < \frac{\pi}{2} \quad (\textit{Pull-Mode}) \\[3mm] \begin{bmatrix} |\mathbf{f}_{\text{ext}}|\cos(\alpha-\beta)\sin(\beta) \\ p_x|\mathbf{f}_{\text{ext}}|\cos(\alpha-\beta)\cos(\beta) \end{bmatrix} & r < r_{\text{in}} \text{ and } |\beta|-|\alpha| > \frac{\pi}{2} \quad (\textit{Push-Mode}) \\[3mm] \begin{bmatrix} 0 \\ 0 \end{bmatrix} & \text{otherwise} \quad (\textit{UR-Mode}) \end{cases}
\tag{7}
$$

with p_x denoting the x-coordinate of the anchor point and the angles α and β as illustrated in Figure 3. The additional conditions that consider the angles α and β in Equation (7) ensure that only forces in the desired direction, based on the actual mode, are projected to the base (e.g., no pushing of the base in *Pull-Mode*). The projected force and torque are then transferred to motion as described in Equation (6). The translational motion of the EE in world coordinates that is caused by a translational motion of the base feels natural and as intended when interacting with the robot. In contrast, rotations of the base cause the hand guided EE to push towards a side, which feels unexpected and unnatural, thus this motion must be compensated. The compensation vector is given by $\mathbf{v}_c = \begin{bmatrix} v_{c,x} & v_{c,y} & \mathbf{0}^{\text{T}} \end{bmatrix}^{\text{T}}$ with $v_{c,x}$ and $v_{c,y}$ as the linear velocities of the EE in x and y direction and $\mathbf{0}^{\text{T}}$ a 4×1 zero vector. The components can be determined as

$$
\begin{bmatrix} v_{c,x} \\ v_{c,y} \end{bmatrix} = \begin{bmatrix} -d\,\dot{\theta}\sin(\beta_{\text{mir}}) \\ d\,\dot{\theta}\cos(\beta_{\text{mir}}) \end{bmatrix} .
\tag{8}
$$

3.2. Haptic Feedback

The haptic feedback provided to the user fulfills several purposes. First, whenever the EE leaves the space between the two circles, so *Push-* or *Pull-Mode* is active, a virtual spring force is generated. This provides the naturally expected resistance when pulling or pushing the mobile base. Second, to avoid approaching singular arm configurations. The avoidance of the shoulder singularity is already guaranteed by means of the inner circle. The remaining two causes for a singularity, a fully stretched elbow (joint 3) and a critical wrist configuration (joint 5), are avoided by adding additional virtual feedback wrenches whenever one of these joint-position gets too close to a critical value. The total wrench-vector for haptic feedback

$$
\tilde{\mathbf{w}}_{\text{fb}} = \tilde{\mathbf{w}}_{\text{s}} + \tilde{\mathbf{w}}_{\text{q}_3} + \tilde{\mathbf{w}}_{\text{q}_5}
\tag{9}
$$

is determined as the sum of the wrench $\tilde{\mathbf{w}}_{\text{s}}$ including the virtual spring forces in *Pull-* or *Push-Mode* and $\tilde{\mathbf{w}}_{\text{q}_3}$ and $\tilde{\mathbf{w}}_{\text{q}_5}$ for singularity-avoidance in joints 3 and 5, respectively.

3.2.1. Virtual Spring

The borders between the three different zones are defined as circles in the xy-plane as shown in Figure 3, resulting in cylindrical shapes in 3D-space, since the z-coordinate of the EE is not taken into account here. Thus, also the virtual spring force acts in the xy-plane only, consequently $\tilde{\mathbf{w}}_s = \begin{bmatrix} F_{s,x} & F_{s,y} & \mathbf{0}^T \end{bmatrix}^T$, where $F_{s,x}$ and $F_{s,y}$ are the x and y components, respectively, and $\mathbf{0}$ denotes the 4×1 zero vector. The equations to determine these components are given by

$$\begin{bmatrix} F_{s,x} \\ F_{s,y} \end{bmatrix} = \begin{cases} \begin{bmatrix} -k_{\text{pull}} \cos(\beta)(r - r_o) \\ -k_{\text{pull}} \sin(\beta)(r - r_o) \end{bmatrix} & r > r_o \ \ (\textit{Pull-Mode}) \\[12pt] \begin{bmatrix} -k_{\text{push}} \cos(\beta)(r - r_i) \\ -k_{\text{push}} \sin(\beta)(r - r_i) \end{bmatrix} & r < r_i \ \ (\textit{Push-Mode}) \\[12pt] \begin{bmatrix} 0 \\ 0 \end{bmatrix} & \text{otherwise} \ \ (\textit{UR-Mode}) \end{cases} \tag{10}$$

with k_{pull} and k_{push} as the spring constants of the virtual springs, r_o and r_i as the radii of the inner and outer circles, respectively, r as the xy-distance between O_E and O_M and the angle β as depicted in Figure 3.

3.2.2. Singularity Avoidance

As discussed in Section 2 there are three types of singularities: The shoulder singularity, the elbow singularity and the wrist singularity. The shoulder singularity is already avoided with the inner circle. Whenever the EE enters this inner circle, a force pointing in the opposite direction is generated, thus by choosing r_i sufficiently large the point P_{56} (see Section 2) cannot reach the plane spanned by the axis of the first and second joint in the base frame of the serial manipulator, despite applying immensely high forces which assume the user will not do.

With a fully stretched elbow, the EE looses its ability to move further away from its base and the arm is in a singular configuration. We avoid this by applying a force to the EE with direction back to origin of the base of the serial manipulator whenever the elbow (joint 3) get closer than a specified distance to the critical joint position, as depicted in Figure 4. The direction of the force is therefore given by the unit-vector $-\mathbf{e}_{O_E}$, which is the negative normalized translation vector of the EE in Σ_M. The pullback-force is determined as

$$\tilde{\mathbf{f}}_{q_3} = \begin{cases} -\mathbf{e}_{O_E} k_3 (q_3 - t_3) & q_3 > t_3 \\ 0 & \text{otherwise} \end{cases} \tag{11}$$

and its magnitude increases, the more the elbow gets stretched. We do not want any feedback torques here, thus $\tilde{\tau}_{q_3} = 0$. The wrench vector for haptic feedback to avoid the elbow singularity is then given by

$$\tilde{\mathbf{w}}_{q_3} = \begin{bmatrix} \tilde{\mathbf{f}}_{q_3} \\ 0 \end{bmatrix}. \tag{12}$$

The wrist singularity occurs, whenever the second wrist joint (joint 5) approaches the position $k\pi$, $k \in \mathbb{Z}$, causing the rotation axes of the other two wrist joints (joints 4 and 6) being parallel. Similar to the avoidance technique for the elbow singularity, we specify a threshold for the minimum distance to the critical joint position. As shown in Figure 4, when the distance falls below this threshold, a virtual torque in the 5-th joint is generated by means of a torsional spring to prevent coming too close to the singular position. The virtual torque is determined as

$$\tau_5 = \begin{cases} k_5(q_5 - t_{5,\text{low}}) & q_5 < t_{5,\text{low}} \\ k_5(q_5 - t_{5,\text{hi}}) & q_5 > t_{5,\text{hi}} \\ 0 & \text{otherwise} \end{cases} \tag{13}$$

where τ_5 is the torque caused by the virtual spring, k_5 denotes the stiffnesses of the virtual torsional spring, q_5 is the angular position of the joint and $t_{i,\text{hi}}$ and $t_{i,\text{low}}$ are the upper and lower thresholds for the virtual spring to become active.

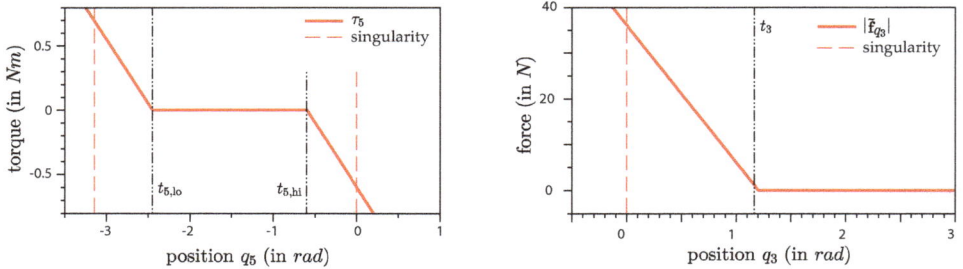

Figure 4. Virtual values for singularity-avoidance: Left: Virtual torque in joint 5. Right: Virtual pullback-force caused by the elbow joint (joint 3).

This virtual torque in the 5-th joint has to be transformed to an associated EE wrench. To determine the reactive force we need the Jacobian $\mathbf{J}$, which is a function of the joint positions $\mathbf{q}_{\text{ur}}$ and composed of a linear part $\mathbf{J}_v$ and a rotational part $\mathbf{J}_\omega$, consequently

$$\mathbf{J} = \begin{bmatrix} \mathbf{J}_v \\ \mathbf{J}_\omega \end{bmatrix} = \begin{bmatrix} j_{v,q_1} & j_{v,q_2} & j_{v,q_3} & j_{v,q_4} & j_{v,q_5} & j_{v,q_6} \\ j_{\omega,q_1} & j_{\omega,q_2} & j_{\omega,q_3} & j_{\omega,q_4} & j_{\omega,q_5} & j_{\omega,q_6} \end{bmatrix}. \tag{14}$$

With the Jacobian we can determine the EE-velocities for a given set of joint-speeds. In particular, we are interested in the linear EE-velocities caused by 5-th joint, which is given in j_{v,q_5}. The reactive force at the EE, caused by a given torque around the axis of rotation of the 5-th joint is indirectly proportional to the distance $|j_{v,q_5}|$, thus we need to invert the magnitude of this vector while maintaining the same direction. This resulting vector is also known as the Samelson inverse and the reactive force is determined as:

$$\tilde{\mathbf{f}}_{q_5} = \frac{j_{v,q_5}}{|j_{v,q_5}|^2} \tau_5 \tag{15}$$

To achieve the desired motion around this axis, the chosen damping coefficients of our controller

$$\mathbf{B}_{\text{ur}} = \begin{bmatrix} \mathbf{B}_v & 0 \\ 0 & \mathbf{B}_\omega \end{bmatrix} \tag{16}$$

need to be taken into account. As given in Equation (6), without external forces ($\mathbf{f}_{\text{ext}} = 0$), the linear velocity-vector $\mathbf{v}$ of the EE, as a reaction to the virtual force $\tilde{\mathbf{f}}_{q_5}$ is given by

$$\mathbf{v} = \mathbf{B}_v^{-1} \tilde{\mathbf{f}}_{q_5}. \tag{17}$$

To Keep the EE on the desired circular trajectory around the axis of rotation of joint 5, the relation between linear and angular velocities

$$\mathbf{v} = |j_{v,q_5}| \boldsymbol{\omega} \tag{18}$$

must hold. The angular EE-velocities are determined by the controller as

$$\omega = \mathbf{B}_\omega^{-1}\tilde{\tau}_{q_5} \tag{19}$$

and thus, to satisfy the constraint from Equation (18), the feedback-torque at the EE is given with

$$\tilde{\tau}_{q_5} = \frac{1}{|\mathbf{j}_{v,q_5}|}\mathbf{B}_\omega\mathbf{B}_v^{-1}\tilde{\mathbf{f}}_{q_5}. \tag{20}$$

The wrench-vector for the haptic feedback of the virtual torsional spring in joint q_5 is given by

$$\tilde{\mathbf{w}}_{q_5} = \begin{bmatrix} \tilde{\mathbf{f}}_{q_5} \\ \tilde{\tau}_{q_5} \end{bmatrix}. \tag{21}$$

4. Experimental Results

To show the effectiveness of the proposed control structure several laboratory experiments were carried out (see supplementary video). This includes straight pulling (Section 4.1) and pushing (Section 4.2) manoeuvres of the EE to demonstrate the working principal of the motion-distribution between serial manipulator and mobile base. A curved pulling experiment (Section 4.3) shows that the mobile manipulator behaves similarly to a simple steered trailer, which we used as inspiration for the controller design. We also show detailed results of the singularity avoidance techniques. As mentioned in Section 3.2.2, the shoulder singularity is avoided by means of the virtual spring of the inner circle. Even tough this is a restrictive choice and permits a large area of the workspace of the serial manipulator it prevents the arm from approaching the shoulder-singularity and no explicit experiments were performed for this case. Results for avoiding the elbow and wrist singularities are discussed in Sections 4.4 and 4.5, respectively. The threshold values $t_3, t_{5,\mathrm{lo}}, t_{5,\mathrm{hi}}$, the elements of the damping matrices $\mathbf{B}_v$, $\mathbf{B}_\omega$, $\mathbf{B}_{\mathrm{mir}}$ as well as the parameters k_{pull}, k_{push}, k_3, k_5 were determined empirically. All parameters used for the experiments are given in Table 1.

Table 1. Table of parameters

Symbol	Value	Unit	Description
r_i	0.48	m	Radius of inner circle
r_o	0.8	m	Radius of outer circle
AP	$\begin{bmatrix} -0.28 & 0 & 0.6 \end{bmatrix}$	m	Anchor-point in Σ_B
$\mathbf{B}_v$	$\begin{bmatrix} 40 & 0 & 0 \\ 0 & 40 & 0 \\ 0 & 0 & 40 \end{bmatrix}$	N·s/m	Translational damping matrix
$\mathbf{B}_\omega$	$\begin{bmatrix} 2 & 0 & 0 \\ 0 & 2 & 0 \\ 0 & 0 & 2 \end{bmatrix}$	Nm·s/rad	Rotational damping matrix
$\mathbf{B}_{\mathrm{mir}}$	$\begin{bmatrix} 50 & 0 & 0 \\ 0 & 1 & 0 \\ 0 & 0 & 7 \end{bmatrix}$	-	Mobile base damping matrix
k_{pull}	140	N/m	Virt. spring stiffness *Pull-Mode*
k_{push}	300	N/m	Virt. spring stiffness *Push-Mode*
k_3	30	-	Constant for pushback-force
k_5	1	N/rad	Virt. spring stiffness in joint 5
t_3	1.2	rad	Position threshold for joint 3
$t_{5,\mathrm{lo}}$	−2.45	rad	Position threshold for joint 5
$t_{5,\mathrm{hi}}$	−0.6	rad	Position threshold for joint 5
$\mathbf{a}_{\mathrm{dh}}$	$\begin{bmatrix} 0 & 0.6127 & 0.5716 & 0 & 0 & 0 \end{bmatrix}$	m	DH-Parameters of UR-10: a
$\mathbf{d}_{\mathrm{dh}}$	$\begin{bmatrix} 0.118 & 0 & 0 & 0.163941 & 0.1157 & 0.0922 \end{bmatrix}$	m	DH-Parameter of UR-10: d
$\boldsymbol{\alpha}_{\mathrm{dh}}$	$\begin{bmatrix} \pi/2 & 0 & 0 & \pi/2 & -\pi/2 & 0 \end{bmatrix}$	rad	DH-Parameter of UR-10: α

4.1. Straight Pulling

The results of straight pulling manoeuvre are shown in Figure 5. The EE starts between the two circles and the controller is in *UR-Mode*, thus the applied force $\mathbf{f}_{ext}$ at the EE initially only causes a motion of the EE. As the radius r increases and the EE leaves the outer circle (first vertical green line), a switch to *Pull-Mode* arises and the applied forces are also projected to the mobile base and cause motion. Withing this experiment, the EE was tried to pull along the negative x-axis, thus the angle β was very small (See Figure 3). As a result, the magnitudes of the projected torque τ_{mir} and the angular velocity $\dot{\theta}$ of the mobile base are small. Once no more force is applied and the EE is released (second vertical green line) the base stops and the EE moves back inside the outer circle.

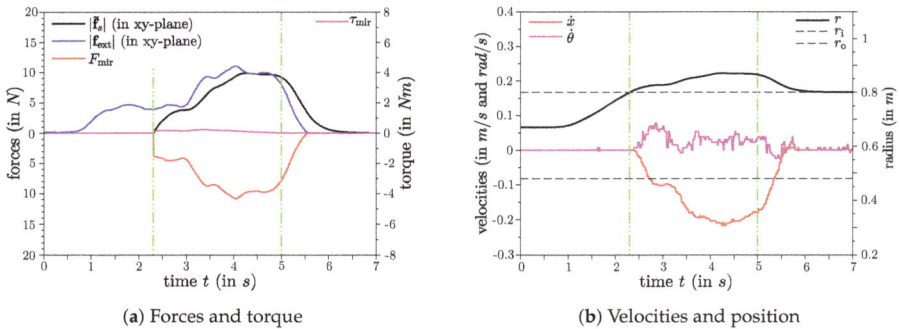

(a) Forces and torque   (b) Velocities and position

Figure 5. Results of a straight pulling manoeuvre: Plot (a): Left axis includes the norm of the external virtual spring torque vector $|\tilde{\mathbf{f}}_s|$ (black), the norm of the external force $|\mathbf{f}_{ext}|$ (blue) and the projected force for the mobile base F_{mir} (red). Right axis shows the projected torque for the mobile base τ_{mir} (magenta). Plot (b): Left axis includes the linear velocity of the mobile base $\dot{x}$ (red) and its rotational velocity $\dot{\theta}$ (magenta). Right axis shows the radius r, which is the xy-distance between EE and UR10 base (black solid) and the radii r_i and r_o of the inner and outer circles (dashed black), respectively.

4.2. Straight Pushing

In Figure 6, the results of a straight pushing manoeuvre are shown. In this experiment, the EE is pushed along the x-axis towards the anchor point. Similar to the straight pulling experiment, the EE starts between the two circles and within the first few seconds only the robotic arm moves until the EE enters the inner circle (first vertical green line). The user receives haptic feedback by the means of the virtual spring with increasing magnitude the deeper the EE enters the inner circle. At the same time, a force and a torque are projected to the mobile base and causes motion there. We tried to push the EE along the x-axis, thus also here the magnitudes of the projected torque and angular velocity of the mobile base are relatively low compared to the curved pulling experiment. As Soon as the EE is released it returns to the inner circle and the mobile base stops.

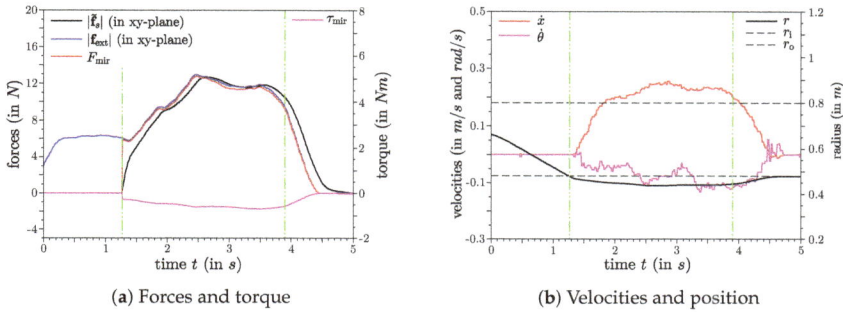

(**a**) Forces and torque

(**b**) Velocities and position

Figure 6. Results of a straight pushing manoeuvre.

4.3. Curved Pulling

For this experiment, a curved pulling action is performed with the results shown in Figure 7. In contrast to the last two experiments, where pulling or pushing happened along the x-axis ($\beta \approx 0$), the EE is pulled with an angle, so that a higher projected torque is generated once the EE leaves the outer circle. This torque causes an angular velocity of the base so that is turns towards the pulling direction. The amplitude of the rotational velocity decreases the closer the EE gets towards the negative x-axis again. Once the base faces the direction only the translational motion remains until releasing the EE.

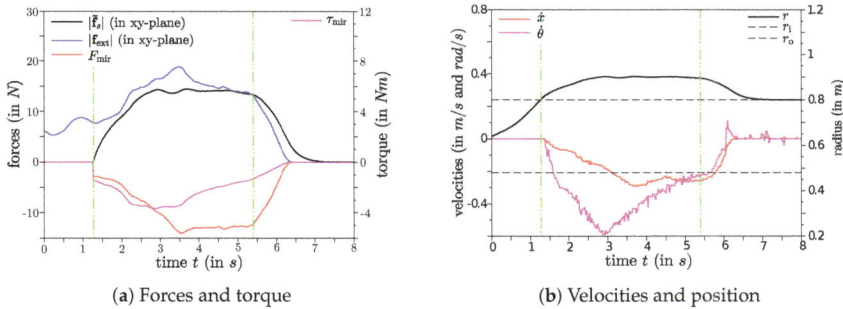

(**a**) Forces and torque

(**b**) Velocities and position

Figure 7. Results of a curved pulling manoeuvre.

4.4. Singularity Avoidance–Elbow-Joint

To show the effectiveness of the proposed technique to avoid a singular configuration caused by the elbow joint, two similar experiments were performed: One with active singularity-avoidance, shown in Figure 8a, and a second with inactive singularity-avoidance ($|\tilde{\mathbf{f}}_{q_3}| = 0$), shown in Figure 8b. For this experiment, the EE starts between the two circles (*UR-mode*) and is pulled upwards. The inner and outer circles are defined in the xy-plane, which results in cylindrical borders in the 3D-space. Without singularity-avoidance it is possible to move the EE in between these cylindrical borders freely, so there is no limitation on the height. This could result in a fully stretched elbow causing a singular arm configuration as demonstrated in Figure 8b (second green line). Please note that within this second experiment no pullback-force is applied when q_3 falls below the threshold value t_3. As a result to the applied pulling-force the elbow stretches more and more until it hits the critical position and the UR10-controller goes into protective stop. The results also show an increasing joint velocity $\dot{q}_3$ as q_3 gets closer to the critical position. This fast joint movement could be very dangerous for humans near the robot and must be avoided. With active singularity-avoidance a pullback-force is applied after the threshold is hit (first green line in Figure 8a) preventing q_3 getting close to the critical position.

During our experiments, it was not possible to get a fully stretched elbow even when excessively high pulling-forces were applied by the user. The working principal of this technique is also depicted in Figure 9.

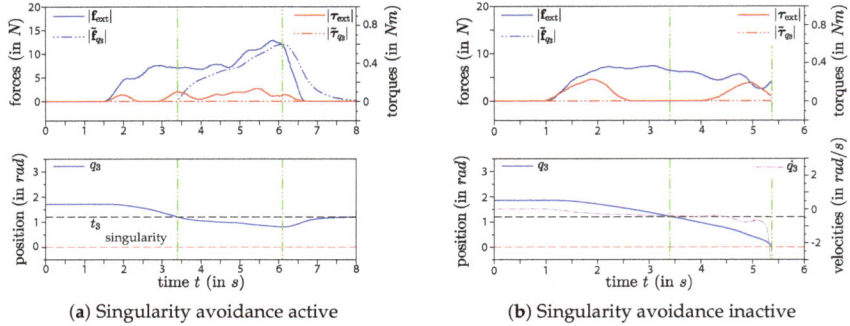

(a) Singularity avoidance active   (b) Singularity avoidance inactive

Figure 8. Elbow-singularity avoidance. Upper plots: Left axis include the norms of the external force $|\mathbf{f}_{ext}|$ (blue) and of the virtual pullback-force $|\tilde{\mathbf{f}}_{q_3}|$ (blue dash-dotted). Right axis include the norms of the external torque $|\boldsymbol{\tau}_{ext}|$ (red) and of the virtual torque $|\tilde{\boldsymbol{\tau}}_{q_3}|$ (red dash-dotted). The lower plots show the joint position q_3 (blue), the position-threshold t_3 (black dashed) and the singular position (red dashed) on the left axis. The lower right plot also shows the joint velocity $\dot{q}_3$ (magenta dash-dotted) on the right axis.

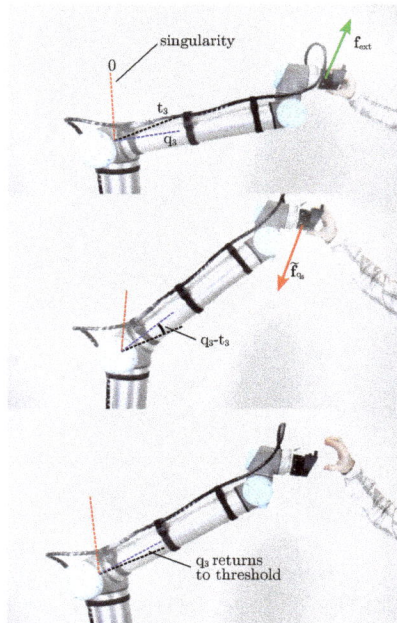

Figure 9. Singularity avoidance in the 3rd joint. The EE is pulled away from the base (top image) causing the elbow joint to move towards the stretched position. As soon as the joint position surpasses the specified threshold t_3 a virtual force is applied to the EE pointing back to the base (middle image). This force increases the more the elbow stretches. Thus, without applying extremely high forces it is not possible for user to get into the singular position of the 3rd joint. After releasing the EE, is moves back towards the base until the position q_3 reaches the threshold (bottom image).

4.5. Singularity Avoidance - Wrist-Joint

The results of the proposed singularity-avoidance strategy for the wrist joint (joint 5) are shown in Figure 10a. For comparison, a similar experiment with inactive singularity-avoidance was carried out with the results shown in Figure 10b. During this experiments, the robot started in a configuration with the joint position q_5 near the upper threshold $t_{q_5,hi}$ and the EE was pushed back towards the base, as depicted in Figure 11. With active singularity avoidance, a virtual force $\tilde{\mathbf{f}}_{q_5}$ and a virtual torque $\tilde{\tau}_{q_5}$ are applied to the EE after q_5 surpasses the threshold (first vertical green line in Figure 10a). As the upper plot shows, the applied and virtual torques have almost the same magnitude. For better readability, only the norms of these vectors are plotted, but these torques are around the same axis but in opposite direction. Thus, they cancel each other in terms of EE motion generation in our controller equation given in Equation (6). Consequently the 5th joint is prevented from rotation further towards the critical position. When the EE is released joint 5 moves back to the threshold. When the singularity avoidance is turned off, the same manoeuvre results in further rotation of the 5th joint towards the singular position, as shown in Figure 10b. This plot also shows that the joint velocities drastically increase when q_5 gets near the critical position (second vertical green line in Figure 10b) which should be avoided in any case.

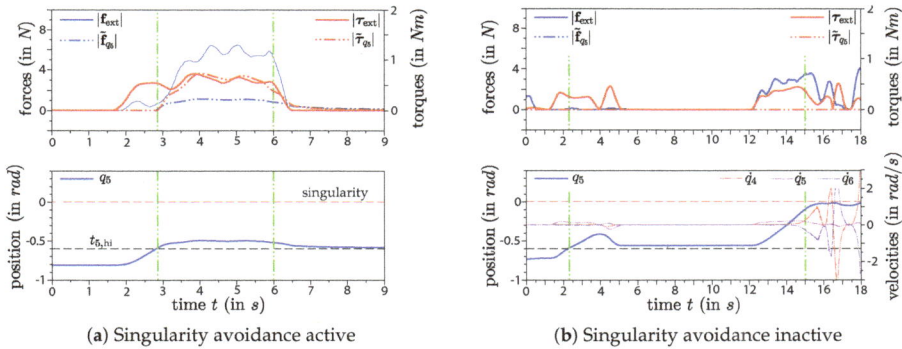

(a) Singularity avoidance active   (b) Singularity avoidance inactive

Figure 10. Wrist-singularity avoidance. Upper plots: Left axis include the norms of the external force $|\mathbf{f}_{ext}|$ (blue) and of the virtual force $|\tilde{\mathbf{f}}_{q_5}|$ (blue dash-dotted). Right axis include the norm of the external torque $|\tau_{ext}|$ (red) and of the virtual torque $|\tilde{\tau}_{q_5}|$ (red dash-dotted). The lower plots show the joint position q_5 (blue), the position-threshold $t_{5,hi}$ (black dashed) and the singular position (red dashed) on the left axis. The lower right plot also shows the joint velocities $\dot{q}_4$, $\dot{q}_5$ and $\dot{q}_6$ (red, magenta and blue dash-dotted) on the right axis.

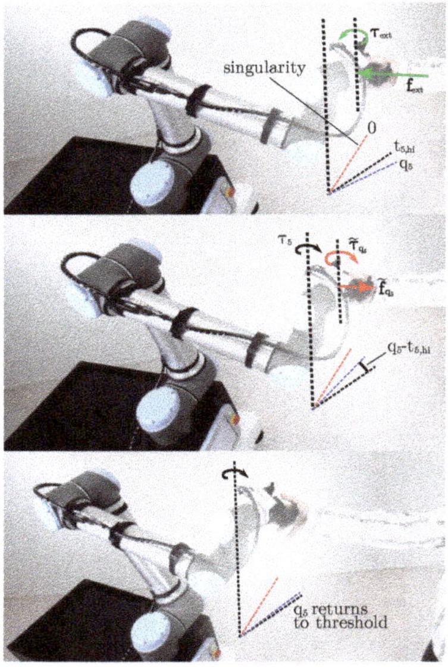

Figure 11. Singularity avoidance in the 5th joint. During this manoeuvre, the EE is pushed towards the mobile base. The User applies the external force $\mathbf{f}_{ext}$ and torque $\boldsymbol{\tau}_{ext}$. This causes a linear and angular motion of the EE. As soon as q_5 surpasses the threshold, a virtual torque and the corresponding virtual wrench at the EE are computed (middle image). As also described in Section 4.5, the external torque and the virtual torque for singularity avoidance cancel each other out and there is no more rotational motion. A translational motion towards the center remains, but it is not possible for user to get into the singular position of the 5th joint.

5. Conclusions

In this work, the practical use of a mobile manipulator was studied and demonstrated. We gave a detailed analysis of all possible singularities for the whole UR robot family and specifically pointed out those of the UR10. We proposed a control structure for hand-guiding the EE in Cartesian coordinates while handling both, the kinematic redundancies of the mobile manipulator and singular configurations of the robot arm. The conducted laboratory experiments on our mobile manipulator CHIMERA show that the system robustly permits these critical arm configuration while allowing the user to guide the EE to the desired target. It is also possible to either move the whole mobile manipulator or only the arm with fixed position of the mobile base without the need for any buttons or additional user interfaces. Moreover, the haptic feedback provided to the user by means of virtual forces and torques makes the interaction very intuitive and easy also for inexperienced users. This system design enables intuitive programming of mobile manipulator tasks using the Programming by Demonstration technique. Additionally the robot can be used as an assistant system without limitations on the workspace, e.g., for gravity compensation tasks. While investigations of the elbow and wrist singularities are straight forward, because each of them solely depends on one particular joint position, analyzing the shoulder singularity is more complex. We showed that our system avoids this configuration, but in a restrictive way since we deny a relatively large area of the manipulator's workspace.

For future work, we plan to refine the avoidance strategy especially for the shoulder singularity. By specifying a metric for the distance to the singularity the volume of the denied workspace could be reduced. Moreover, there are multiple solutions for the inverse kinematics of the serial manipulator. Switching from one posture the another implies going through a singularity and the current system design does not allow for manually switching the configuration (e.g., from elbow-up to elbow-down).

Robotics **2019**, *8*, 14

A singularity transition strategy could therefore also be useful to overcome this issue. Furthermore, we plan to eliminate the force-torque sensor on the EE. This means that we use the estimated external wrench based on the joint sensor values instead.

Supplementary Materials: The following are available online at www.mdpi.com/2218-6581/8/1/14/s1, a Video of the conducted experiments is included.

Author Contributions: M.W. and M.B. conceived and designed the control structure. M.H. performed the singularity analysis. M.W. performed the implementation on the robot and the experiments. M.W. and M.B. analyzed the data. All authors contributed to the writing process.

Funding: This research was funded by the Austrian Ministry for Transport, Innovation and Technology (BMVIT) within the framework of the sponsorship agreement formed for 2015-2018 under the project RedRobCo.

Conflicts of Interest: The authors declare no conflict of interest.

References

1. Pai, Y.S.; Yap, H.J.; Singh, R. Augmented reality—Based programming, planning and simulation of a robotic work cell. *J. Eng. Manuf.* **2015**, *229*, 1029–1045. [CrossRef]
2. Van den Bergh, M.; Carton, D.; De Nijs, R.; Mitsou, N.; Landsiedel, C.; Kuehnlenz, K.; Wollherr, D.; Van Gool, L.; Buss, M. Real-time 3D hand gesture interaction with a robot for understanding directions from humans. In Proceedings of the International Conference on Robot and Human Interactive Communication (RO-MAN), Atlanta, GA, USA, 31 July–3 August 2011; pp. 357–362.
3. Valner, R.; Kruusamäe, K.; Pryor, M. TeMoto: Intuitive Multi-Range Telerobotic System with Natural Gestural and Verbal Instruction Interface. *Robotics* **2018**, *7*, 9. [CrossRef]
4. Goto, S. Forcefree control for flexible motion of industrial articulated robot arms. In *Industrial Robotics: Theory, Modelling and Control*; Pro Literatur Verlag: Mammendorf, Germany, 2006.
5. Akgun, B.; Cakmak, M.; Yoo, J.W.; Thomaz, A.L. Trajectories and keyframes for kinesthetic teaching: A human-robot interaction perspective. In Proceedings of the seventh annual ACM/IEEE international conference on Human-Robot Interaction, Boston, MA, USA, 5–8 March 2012; pp. 391–398.
6. Elliott, S.; Toris, R.; Cakmak, M. Efficient programming of manipulation tasks by demonstration and adaptation. In Proceedings of the 26th IEEE International Symposium on Robot and Human Interactive Communication (RO-MAN), Lisbon, Portugal, 28 August–1 September 2017; pp. 1146–1153.
7. Villani, V.; Pini, F.; Leali, F.; Secchi, C. Survey on human–robot collaboration in industrial settings: Safety, intuitive interfaces and applications. *Mechatronics* **2018**, *55*, 248–266. [CrossRef]
8. Vischer, D.; Khatib, O. Design and development of high-performance torque-controlled joints. *Trans. Robot. Autom.* **1995**, *11*, 537–544. [CrossRef]
9. Bicchi, A.; Rizzini, S.L.; Tonietti, G. Compliant design for intrinsic safety: General issues and preliminary design. In Proceedings of the IEEE/RSJ International Conference on Intelligent Robots and Systems, Maui, HI, USA, 29 October–3 November 2001; pp. 1864–1869.
10. Hirzinger, G.; Sporer, N.; Albu-Schaffer, A.; Hahnle, M.; Krenn, R.; Pascucci, A.; Schedl, M. DLR's torque-controlled light weight robot III-are we reaching the technological limits now? In Proceedings of the International Conference on Robotics and Automation (ICRA), Washington, DC, USA, 11–15 May 2002; pp. 1710–1716.
11. Zollo, L.; Siciliano, B.; Laschi, C.; Teti, G.; Dario, P. An experimental study on compliance control for a redundant personal robot arm. *Rob. Autom. Syst.* **2003**, *44*, 101–129. [CrossRef]
12. Zinn, M.; Roth, B.; Khatib, O.; Salisbury, J.K. A new actuation approach for human friendly robot design. *Int. J. Robot. Res.* **2004**, *23*, 379–398. [CrossRef]
13. Haddadin, S.; Huber, F.; Albu-Schäffer, A. Optimal control for exploiting the natural dynamics of variable stiffness robots. In Proceedings of the IEEE International Conference on Robotics and Automation (ICRA), Saint Paul, MN, USA, 14–18 May 2012; pp. 3347–3354.
14. Leboutet, Q.; Dean-León, E.; Cheng, G. Tactile-based compliance with hierarchical force propagation for omnidirectional mobile manipulators. In Proceedings of the IEEE-RAS 16th International Conference on Humanoid Robots (Humanoids 2016), Cancun, Mexico, 15–17 November 2016; pp. 926–931.

15. Navarro, B.; Cherubini, A.; Fonte, A.; Poisson, G.; Fraisse, P. A framework for intuitive collaboration with a mobile manipulator. In Proceedings of the IEEE/RSJ International Conference on Intelligent Robots and Systems (IROS), Vancouver, BC, Canada, 24–28 September 2017; pp. 6293–6298.
16. Han, H.; Park, J. Robot control near singularity and joint limit using a continuous task transition algorithm. *Int. J. Adv. Robot. Syst.* **2013**, *10*, 346. [CrossRef]
17. Weyrer, M.; Brandstötter, M.; Mirkovic, D. Intuitive Hand Guidance of a Force-Controlled Sensitive Mobile Manipulator. In Proceedings of the IFToMM Symposium on Mechanism Design for Robotics, Udine, Italy, 11–13 Augus 2018; pp. 361–368.
18. Siegwart, R.; Nourbakhsh, I.R.; Scaramuzza, D.; Arkin, R.C. *Introduction to Autonomous Mobile Robots*; MIT Press: Cambridge, MA, USA, 2011.
19. Kebria, P.M.; Al-Wais, S.; Abdi, H.; Nahavandi, S. Kinematic and dynamic modelling of UR5 manipulator. In Proceedings of the IEEE International Conference on Systems, Man, and Cybernetics (SMC), Budapest, Hungary, 9–12 October 2016.
20. Husty, M.L.; Karger, A.; Sachs, H.; Steinhilper, W. *Kinematik und Robotik*, 1st ed.; Springer: New York, NY, USA, 1997.
21. Siciliano, B.; Sciavicco, L.; Villani, L.; Oriolo, G. *Robotics: Modelling, Planning and Control*; Springer: New York, NY, USA, 2010.
22. Albu-Schäffer, A.; Ott, C.; Hirzinger, G. A unified passivity-based control framework for position, torque and impedance control of flexible joint robots. *Int. J. Robot. Res.* **2007**, *26*, 23–39. [CrossRef]
23. Wahrburg, A.; Bös, J.; Listmann, K.D.; Dai, F.; Matthias, B.; Ding, H. Motor-Current-Based Estimation of Cartesian Contact Forces and Torques for Robotic Manipulators and Its Application to Force Control. *IEEE Trans. Autom. Sci. Eng.* **2018**, *15*, 879–886. [CrossRef]

robotics

MDPI

Article

Trajectory Design for Energy Savings in Redundant Robotic Cells †

Paolo Boscariol * and Dario Richiedei

Dipartimento di Tecnica e Gestione dei Sistemi Industriali, Università degli Studi di Padova,
36100 Vicenza, Italy; dario.richiedei@unipd.it
* Correspondence: paolo.boscariol@unipd.it
† This paper is an extended version of a conference paper previously published In the Proceedings of the 4th
IFToMM Symposium on Mechanism Design for Robotics, Udine, Italy, 11–13 September 2018, "Energy
Saving in Redundant Robotic Cells: Optimal Trajectory Planning".

Received: 17 January 2019; Accepted: 16 February 2019; Published: 20 February 2019

Abstract: This work explores the possibility of exploiting kinematic redundancy as a tool to enhance the energetic performance of a robotic cell. The test case under consideration comprises a three-degree-of-freedom Selective Compliance Assembly Robot Arm (SCARA) robot and an additional linear unit that is used to move the workpiece during a pick and place operation. The trajectory design is based on a spline interpolation of a sequence of via-points: The corresponding motion of the joints is used to evaluate, through the use of an inverse dynamic model, the actuators effort and the associated power consumption by the robot and by the linear unit. Numerical results confirm that the suggested method can improve both the execution time and the overall energetic efficiency of the cell.

Keywords: trajectory planning; energy efficiency; redundancy; robotic cell; kinematic redundancy

1. Introduction

The efficient use of resources is one of the challenges that industry has to face, not only to reduce the manufacturing and handling costs, but also to comply with the directives set by the European Union [1,2]. Such directives, which encourage the adoption of energy efficiency improvements, provide a strong market pull for energy-efficiency enabling technologies and research activities. The latter are testified by the flourishing literature [3] on the topic. Some recent theoretical [4] and experimental [5] investigations have shown that software and hardware solutions can lead to up to a 30% energetic improvement for robotic systems.

The work [3] proposes a classification of the numerous solutions proposed for the energetic improvement of automatic and and robotic systems. The classification identifies lightweight robot design [6], energy recovery and storage [7], robot architecture selection [8] and motion planning [9,10] as the main tools for achieving sensible energy consumption mitigation of automatic machines. Facing the problem of enhancing energetic efficiency by focusing on motion design has the advantage of being a 'software solution' that can be applied to a wide range of systems. Not only are industrial robots the subject of studies on energy efficiency, since energy-saving trajectories for mobile robots are under investigation as well [11,12].

The impact of motion planning on the energy consumption of electric-driven mechatronic devices is well known and well understood, having been analyzed since the 1970s [13]. Several works have shown that simple analytical models can be effectively used to evaluate and optimize the power consumption of automatic machines [14,15]. These models can be used to formulate optimization problems that can be solved analytically for simpler cases or numerically for more complex ones [16].

Another tool that can be effectively exploited when designing energy-efficient motion profiles is kinematic redundancy, i.e., the availability of extra degrees of freedom of the robotic system [17]. Redundancy provides additional degrees of freedom to the solution of the inverse kinematic problem as well as to the choice of the motion profiles of the extra degrees of freedom. Increasing the number of degrees of freedom allows also to define a more energy efficient torque distribution among the actuators, as demonstrated for parallel robots in [18–21]. The work [18] analyzes several possible modifications to a planar kinematic robot, with the aim of finding the one that guarantees the higher energy savings, under a torque distribution managed by a predictive strategy. The numerical results are then extended to a full experimental validation in [19]. The optimal torque distribution in a redundantly actuated system can also be obtained by off-line optimization routines, as suggested in [20] or in [21].

This work suggest a novel approach to the topic of energy efficient operation of robots by discussing the optimization of the motion profiles for kinematically redundant robotic cells. This works proposes, as a test case, a robotic cell made by a three degrees of freedom (DOFs) Selective Compliance Assembly Robot Arm (SCARA) robot and a linear unit, which can be used to move the workpiece. Instead of introducing permanent modifications to the kinematics of an existing machine, as suggested in [18–21], here redundancy is introduced by adding an additional external and independent axis to a standard robot. A numerical optimization tool is suggested to fully exploit the availability of the extra degree of freedom through a careful design of the robot trajectory and of the motion design of the additional linear unit. The energy saving is then evaluated by comparing the result of the application of the proposed method with and without the use of the additional linear unit.

2. Energy Consumption Estimation

In this section the analytical models used for evaluating the energy consumption of the robotic cell is recalled. The goal is to find an expression for the energy consumption associated with a robotic task, which in this case is expressed as a trajectory in the operative space. Two models are presented, to describe separately the SCARA robot and the linear unit. For the second one, owing to its simpler dynamics, an analytic closed-form expression of the energy consumption will be presented.

2.1. SCARA Robot

The robotic cell used as a testbench comprises a three DOFs SCARA robot, as shown in Figure 1.

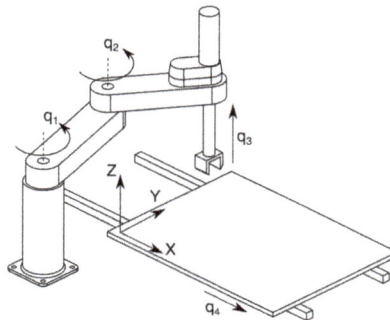

Figure 1. Layout of the robotic cell.

The SCARA robot is actuated by three brushless motors, and its main electric and mechanical parameters are reported in Table 1. The dynamic model of the SCARA robot can be obtained by using the Lagrangian formalism, leading to the usual formulation:

Table 1. Mechanical and electric parameters of the Selective Compliance Assembly Robot Arm (SCARA) robot.

Parameter	Joint 1	Joint 2	Joint 3
Link length	0.45 m	0.35 m	-
Link mass	14 kg	18 kg	2 kg
Gear ratio	1/30	1/30	1/30
Motor inertia	1×10^{-4} kg m^2	1×10^{-4} kg m^2	1×10^{-4} kg m^2
Viscous friction coefficient	0.001 Nm s/rad	0.001 Nm s/rad	0.001 Nm s / rad
Coulomb friction force	2×10^{-3} Nm	2×10^{-3} Nm	2×10^{-3} Nm
Motor winding resistance	3 Ω	3 Ω	3.5 Ω
Motor back-emf constant	0.6 Vs/rad	0.6 Vs/rad	0.6 Vs/rad
Motor torque constant	0.6 Nm/A	0.6 Nm/A	0.6 Nm/A
Peak motor torque	2.5 N	2.5 N	1.5 N
Peak motor power	75 W	75 W	50 W

$$\mathbf{M}(\mathbf{q})\ddot{\mathbf{q}} + \mathbf{C}(\mathbf{q}, \dot{\mathbf{q}}) + \mathbf{f}_v \dot{\mathbf{q}} + \mathbf{F}_c \, \text{sign}\,(\dot{\mathbf{q}}) = \boldsymbol{\tau}_m \tag{1}$$

Equation (1) involves the vector of joint coordinates $\mathbf{q} = [q_1, q_2, q_3]^T$, the diagonal matrix $\mathbf{f}_v$ of viscous friction coefficients and the diagonal matrix of Coulomb friction forces $\mathbf{F}_c$. Motor torques at the joint are represented by the three components of vector $\boldsymbol{\tau}_m$. $\mathbf{M}(\mathbf{q})$ is the configuration-dependent mass matrix, while $\mathbf{C}(\mathbf{q}, \dot{\mathbf{q}})$ accounts for the centrifugal effects. The electromechanical model of the motors that drive the SCARA robot can be introduced into Equation (1), by recalling that the motor currents and the motor torques can be related by the diagonal matrix of the motor torque constants, $\mathbf{k}_t$, as:

$$\boldsymbol{\tau}_m(t) = \mathbf{k}_t \mathbf{I}(t) \tag{2}$$

The voltage drop across the motors can then be described by the equivalent DC motor armature model, which collects the contributions for all the joints of the robot:

$$\mathbf{V}(t) = \mathbf{R}\mathbf{I}(t) + \mathbf{L}\frac{d\mathbf{I}(t)}{dt} + \mathbf{k}_b \dot{\mathbf{q}}_m(t) \tag{3}$$

where $\dot{\mathbf{q}}_m$ is the vector of the motor velocities, $\mathbf{k}_b$ is the diagonal matrix of the motor back-emf constants, $\mathbf{R}$ is the matrix of motor winding resistances and $\mathbf{L}$ is the inductance matrix. The instantaneous power drawn by the robot is then simply expressed by the current-voltage product:

$$\mathbf{W}_e(t) = \mathbf{V}^T(t)\mathbf{I}(t) \tag{4}$$

If regenerative drives are assumed, and energy loss due to regeneration is neglected, the overall energy consumption for the time interval $[t_a, t_b]$ is found by computing the time integral:

$$E_{SCARA} = \int_{t_a}^{t_b} \mathbf{V}^T(t)\mathbf{I}(t)dt \tag{5}$$

This model can be implemented within the trajectory optimization routine, which will include the numerical integration of Equation (5), following the approach commonly used in works such as [22]. It must be pointed out that the electric power, as expressed in Equation (4), takes positive values when the energy is drawn by the robot from the drive unit and negative values when the energy flow is reversed. Whenever the motor drives does not support regeneration [23], the current drawn from the actuators during the braking phase is dissipated by a braking resistor, and therefore negative values of $\mathbf{P}_e(t)$ should be not accounted for in Equation (5). Both the cases of regenerative and non-regenerative robot drives are taken into consideration in Section 3 to highlight the results of the energy optimization in both cases.

2.2. Linear Unit

The estimation of the energy needed to drive the fourth axis of the robotic cell, i.e., the sliding table, can be efficiently performed using an analytical closed-form formulation rather then by numerical integration, as imposed by the analytical complexity of the SCARA robot dynamics.

The dynamics of the linear unit can be described by a single differential equation, that takes the common form of the dynamics of a constant inertia systems, as:

$$m\ddot{q}_4(t) + f_v\dot{q}_4(t) + F_c \, \text{sign} \, (\dot{q}_4(t)) = F_m(t) \tag{6}$$

The parameters appearing in Equation (6) are the moving mass of the linear unit m, the coefficient of viscous friction f_v and the Coulomb friction force F_c. Since the linear unit is actuated in a direct drive arrangement, i.e., without the use of transmission system, the force acting on the moving mass is equal to the force $F_m(t)$ exerted by the brushless linear motor which drives the unit. If required, this formulation can be extended in a straightforward manner to include reversible transmission system, such as a ball screw mechanism. The values used to define the model of the linear unit are shown in Table 2.

Table 2. Mechanical and electric parameters of the linear unit.

Parameter	Value
Moving mass	5.18 kg
Viscous friction coefficient	1×10^{-3} Ns/m
Coulomb friction force	2×10^{-2} N
Back-emf constant	3.1 Vs/m
Torque constant	3.12 N/A

The equations that describe the electric dynamics of the linear unit are the scalar equivalents of the ones already used for the robot, i.e., Equations (2) and (3). Accordingly, the instantaneous electric power drawn by the linear unit can be written as:

$$W_{LU}(t) = \frac{R}{k_t^2} \left(m^2\ddot{q}_4^2(t) + f_v\dot{q}_4^2(t) + F_c^2 + 2mf_v\dot{q}_4(t)\ddot{q}_4(t) + 2f_vF_c\dot{q}_4(t) \right)$$
$$+ \frac{k_b}{k_t} \left(m\dot{q}_4(t)\ddot{q}_4(t) + f_v\dot{q}_4^2(t) + F_c\dot{q}_4(t) \right) \tag{7}$$

Equation (7) does not include the effects of inductance on power absorption, since they are negligible. Equation (7) shows that the instantaneous electric power drawn by the motor is the sum of a constant term, associated with the Coulomb friction force F_c, and of a term that depends on speed, acceleration, on their product and on their squared values. The time integration of Equation (7) over the interval $[0, T]$ results in the estimation of the energy required to perform a generic motion profile. In a rest-to-rest motion profiles the energy associated with the acceleration $\ddot{q}_4(t)$, with its squared value $\ddot{q}_4^2(t)$ and with the speed-acceleration product are null, i.e.,

$$\int_0^T \ddot{q}_4(t)dt = 0; \qquad \int_0^T \ddot{q}_4^2(t)dt = 0; \qquad \int_0^T \dot{q}_4(t)\ddot{q}_4(t)dt = 0; \tag{8}$$

Furthermore, the power contribution that is directly proportional to the speed, when integrated, is simply related to the overall displacement H, given that:

$$\int_0^T \dot{q}_4(t)dt = H \tag{9}$$

The overall energy consumption can therefore be rewritten by rearranging Equation (7) and by computing its definite integral over $[0, T]$ as:

$$E_{LU} = \left(\frac{R}{k_t^2} + \frac{k_b}{k_t} \right) f_v \int_0^T \dot{q}_4^2(t) dt + \left(2\frac{R}{k_t^2} f_v + \frac{k_b}{k_t} \right) F_c H$$
$$+ F_c \frac{R}{k_t^2} T + \frac{R}{k_t^2} m^2 \int_0^T \ddot{q}_4^2(t) dt \tag{10}$$

The formulation of Equation (10) can also be rearranged by highlighting the contributions proportional to the RMS value of joint speed $\dot{q}_4$ and acceleration $\ddot{q}_4$ using the formulas:

$$\int_0^T \dot{q}_4(t) dt = T \dot{q}_{4,RMS}^2 \tag{11}$$

$$\int_0^T \ddot{q}_4(t) dt = T \ddot{q}_{4,RMS}^2 \tag{12}$$

Accordingly, the energy used by the linear unit can be computed as the sum of four terms, highlighting the proportionality to the total displacement H, to the motion time T, to the root-means-square (RMS) values of speed and acceleration, as:

$$E_{LU} = \left(\frac{R}{k_t^2} + \frac{k_b}{k_t} \right) f_v T \dot{q}_{4,RMS}^2 + \left(2\frac{R}{k_t^2} f_v + \frac{k_b}{k_t} \right) F_c H + F_c \frac{R}{k_t^2} T + \frac{R}{k_t^2} m^2 T \ddot{q}_{4,RMS}^2 \tag{13}$$

Equation (13) is of general application, since it can be used for any choice of the rest-to-rest motion profile of the linear unit, provided that the the RMS values of the table speed and acceleration can be computed. As suggested in [24], their computation can be performed using the characteristic coefficients of RMS speed and acceleration, $c_{V_{RMS}}$ and $c_{A_{RMS}}$, which are commonly available in standard reference books [25]. Such coefficients are computed with reference to a normalized motion profile $q_N(\tau)$, which provides a unitary displacement over a unitary time duration, as:

$$C_{v,RMS} = \sqrt{\int_0^T \left(\frac{dq_N}{d\tau} \right)^2 d\tau} \tag{14}$$

$$C_{a,RMS} = \sqrt{\int_0^T \left(\frac{d^2 q_N}{d\tau^2} \right)^2 d\tau} \tag{15}$$

Recalling that [24]:

$$\dot{q}_{4,RMS} = \frac{H}{T} C_{v,RMS} \tag{16}$$

$$\ddot{q}_{4,RMS} = \frac{H}{T^2} C_{a,RMS} \tag{17}$$

the total energy consumption of the linear unit while performing a generic rest-to-rest motion can be evaluated also as:

$$E_{LU} = \left(\frac{R}{k_t^2} + \frac{k_b}{k_t} \right) f_v T \left(\frac{h}{T} c_{v,RMS} \right)^2 + \left(2\frac{R}{k_t^2} f_v + \frac{k_b}{k_t} \right) F_c H + F_c \frac{R}{k_t^2} T + \frac{R}{k_t^2} m^2 T \left(\frac{h}{T^2} c_{a,RMS} \right)^2 \tag{18}$$

3. Energy Optimization

The energy consumption models reported in Section 2 can be used as the basis of an energy optimization problem. Countless choices are available as far as the trajectory primitive is concerned, according to the extensive results available in literature [25,26]. As a matter of example, in this work the trajectory design is performed with reference to the so-called '434' spline algorithm. This motion profile, which was introduced by Cook and Ho in [27], is based on the use of piecewise polynomial functions as the means to produce a trajectory that passes through a sequence of via-points, which might be expressed either in the operative or in the joint space domain. The motion primitive takes its name from the sequence of third and fourth degree polynomials function that are used to provide the interpolation of the via-points with continuity up to the second derivative, so that continuous acceleration is always achieved. Further alterations of the original definition can be performed to achieve speed, acceleration and jerk limitations [28].

In the case under consideration in this work an energy saving trajectory design is sought, using the 434 trajectory for the SCARA robot, even if the method can be applied, with obvious modifications, with spline algorithms and with arbitrary motion profile of the linear unit. The task is specified by N via-points that are usually defined, for the operator's convenience, in the operative space, meanwhile the computation of the interpolated motion profile according to the aforementioned 434 algorithm is performed after their transformation to the joint space.

This motion profile is used here also to exploit its simple parametrization: Once a sequence of via-points is defined, the trajectory is designed by setting the time distance between each two consecutive via-points. Each trajectory implementation is uniquely identified by a vector of time distances **T**, which comprises $N - 1$ elements, each one identified as T_i. This feature makes the '434' trajectory profile very suitable to be the basis of an optimization procedure.

As far as the optimization of the motion of the linear unit is concerned, here the 'symmetric double-S' profile is chosen as the motion law, but the procedure can be adopted to other motion profiles using a similar procedure. The motion is conveniently described by the piecewise expression of the velocity as:

$$
\begin{aligned}
\dot{q}_4(t) &= \frac{v_0}{2}\left(1 - cos(\omega_1 t)\right); & t &\in [0, t_a) \\
\dot{q}_4(t) &= v_0; & t &\in [t_a, T - t_a) \\
\dot{q}_4(t) &= \frac{v_0}{2}\left(1 + cos(\omega_1 t)\right); & t &\in (T - t_a, T]
\end{aligned}
\tag{19}
$$

with

$$
v_0 = \frac{H}{T - t_a}; \qquad \omega_1 = \frac{\pi}{t_a};
\tag{20}
$$

The motion of the linear unit is therefore split into three phases, comprising an acceleration and deceleration phase, each one lasting t_a seconds, and a constant speed phase. Equation (20) ensures that the whole displacement H is performed in the total time T. If the total motion time T is prescribed, the optimization of the trajectory of the linear unit can be performed by setting the acceleration time t_a within the range $(0, T/2]$. After evaluating the coefficients $c_{v,RMS}$ and $c_{a,RMS}$ according to Equations (19) and (20) and by using them into Equation (18), the energy consumption of the linear unit is:

$$
E_{LU} = f_v \left(\frac{R}{k_t^2} + \frac{k_b}{k_t}\right)\left(T - \frac{5}{4}t_a\right)v_0^2 + \left(\frac{R}{k_t^2}2f_v + \frac{k_b}{k_t}\right)F_cH + F_c\frac{R}{k_t^2}T + \frac{R}{t_a}\left(\frac{mv_0}{2k_t}\right)^2
\tag{21}
$$

Equation (21) is written for the double-S motion profile, but it should be highlighted that the formulation of Equation (18) can be applied to any arbitrary motion profile, simply by finding the corresponding values for the characteristic coefficients $c_{v,RMS}$ and $c_{a,RMS}$.

Once a method for evaluating the energy consumption of the whole cell is established, a numerical optimization can be set up according to the following problem:

$$\min_{[\mathbf{T}, t_a]} E_{SCARA} + E_{LU} \tag{22}$$

$$\text{with: } \mathbf{T} \in R^{N-1};$$

$$t_a \in (0, T/2];$$

$$\text{subject to: } \sum_{i=1}^{N-1} T_i = T;$$

$$\text{bounded } |\dot{\mathbf{q}}|, |\ddot{\mathbf{q}}|, |\dddot{\mathbf{q}}|;$$

$$\text{bounded } |W_{e,j}|, |\tau_j|, \text{ with } j = 1, 2, 3;$$

The cost function used in Equation (22) equally weighs the energy required by the robot and by the linear unit, therefore the optimization problem is targeted at reducing the overall energy consumption of the robotic cell. The energy required by the SCARA robot, E_{SCARA} is evaluated according to Equations (2)–(5), while E_{LU} is evaluated using just Equation (21). Optimization variables includes the set of $N-1$ times T_i between two consecutive via-points, which are collected in the vector $\mathbf{T}$, and the acceleration time t_a, which sets the design of the motion profile of the linear unit according to Equations (19) and (20). The whole task is designed to be performed in T seconds, and the motion of both devices happens simultaneously and without any pauses. In addition to the bounds on absolute values joint speed, acceleration and jerk, the optimization problem in Equations (22) can account for limits on the peak values of the motor torques τ_j and on the electric power $W_{e,j}$ draw by the j-th motor. The last two constraints are enforced to ensure the limitation imposed by the robot drive unit.

The optimization problem is not convex, however a careful selection of the initial guess has shown to be capable of getting rid of this issue and to boost the achievement of significant energy reductions. A sensible initial guess for the starting value of the times T_i can be obtained using either the chord length distribution or the centripetal distribution methods, as proposed in literature [25].

The benchmark problem taken into consideration in this work is described by six via-points defined in a reference frame located on the sliding table. The six via-points, which are reported in Table 3, reproduce a pick & place task that involves the motion of all the four axes of the robotic cell. This choice reproduces a typical task performed by SCARA robots in assembly or packaging applications. The energetic performance of the proposed method is measured in comparison with a non-redundant solution, the latter being obtained without the use of the linear unit, i.e., by enforcing $\dot{q}_4(t) = 0$.

Table 3. Coordinates of the via-points for the pick & place task.

Via-point	X [m]	Y [m]	Z [m]
1	0.6	0.2	0
2	0.6	0.2	0.2
3	0.6	0.2	0.3
4	−0.2	0.4	0.3
5	−0.2	0.4	0.2
6	−0.2	0.4	0

3.1. Trajectory Optimization with Electric Power Regeneration

Figure 2 reports the results of the solution of the optimization problem of Equation (22) for a sequence of execution times varying from $T = 7$ s to $T = 2$ s, halting the iteration when reducing the minimum time allowed by the constraints. The procedure was repeated for the redundant and the non-redundant task.

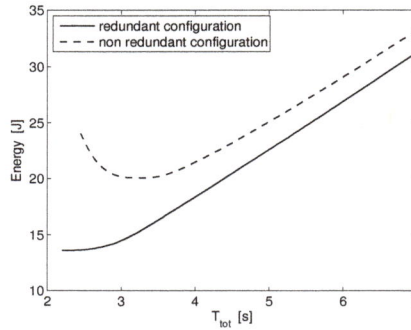

Figure 2. Energy consumption versus total execution time: Comparison between the redundant and non-redundant configuration, with regenerative actuation.

The solid line in Figure 2 shows the minimum energy required to perform the task in the redundant case, while the dashed line refers to the use of the SCARA robot only. In all the cases included in Figure 2 the redundant configuration is by a noticeable amount the most energetically efficient one. It can also be highlighted that the advantage in terms of energy saving allowed by the redundancy is more relevant for faster motion profiles.

The non-redundant configuration requires at best 20.5 J to complete the task, while just 13.6 J are needed when using the linear unit. Optimal execution times are 3.23 s and 2.25 s, respectively. Figures 3 and 4 show the optimal trajectories as designed through the optimization routine.

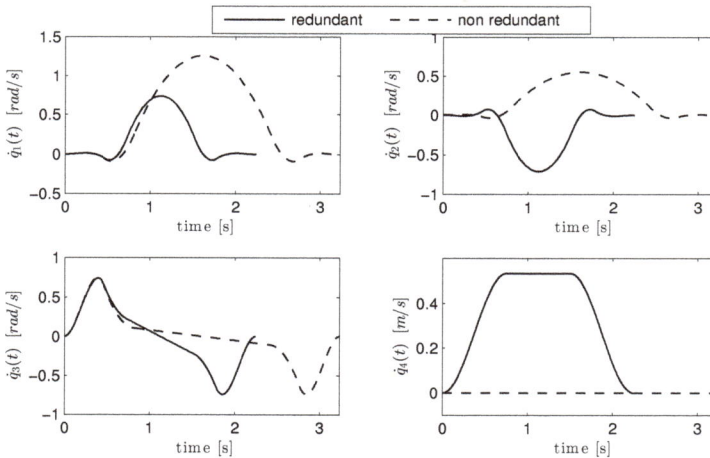

Figure 3. Joint speed: Comparison between the energy-optimal trajectories with redundant and non-redundant configuration.

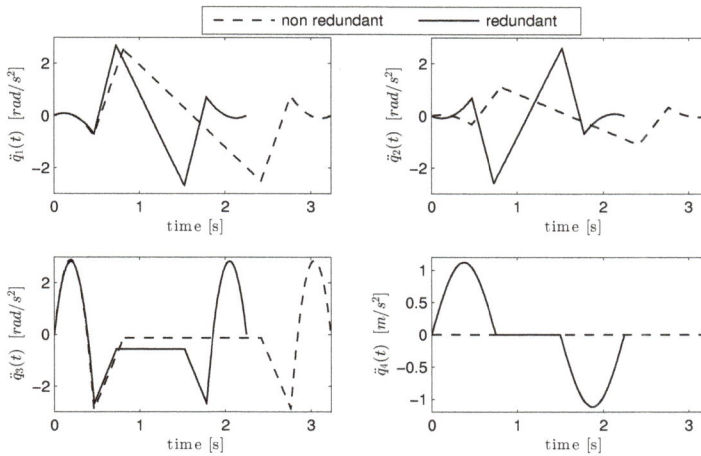

Figure 4. Joint acceleration: Comparison between the energy-optimal trajectories with redundant and non-redundant configuration.

In particular, Figure 3 shows the speed of each axis for the minimum energy solutions. It can be seen that the motion of joints 2 and 3 are radically different when switching from the redundant to the non-redundant configuration, since the motion along the X direction is in one case provided by the SCARA robot, and in the other one is provided by the linear unit. The frame of reference for the Cartesian motion of the end-effector is shown in Figure 1. Hence the SCARA robot performs a smaller displacement in the redundant case. The motion along the vertical direction, which is provided by the third axis, follows a similar profiles in both cases, with the noticeable difference of a sensibly shorter execution time in the redundant case due to the smaller optimal motion duration.

The corresponding paths of the SCARA end effector are shown in Figure 5: In the non-redundant case the motion of the end-effector passes exactly through all the six via-point, while in the redundant configuration just the first and the last via-points are touched by the path. The corresponding distortion of the path is due to the absence of a synchronization between the motion of the robot and of the linear unit, as the result of using a rest-to-rest motion profile for the linear unit and a via-point approach for the robot.

It is indeed worthwhile to notice that for the third joint, the peak speed values are essentially unaffected by the shorter execution time associated with the redundant configuration: Also the corresponding acceleration, as it can be seen in Figure 4, shows almost identical profiles of the first and the last segment of the trajectory. In all cases the peak values of speed, acceleration and, although not shown, jerk, are comfortably within the kinematic bounds imposed when defining the optimization procedure.

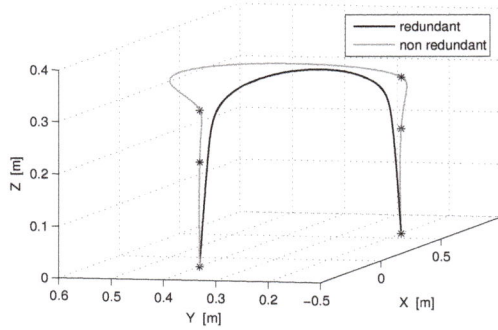

Figure 5. Path in the operative space: Comparison between redundant and non-redundant configuration.

The profiles of the absorbed electric power obtained in the same tests are are reported in Figure 6. The dashed lines, which refer to the non-redundant configuration, shows that all the motors that drive the SCARA robot draw electric energy during the first half of the task, while a fraction of that energy is fed back to motor drives in the remaining part of the trajectory. The power profiles optimized for the redundant trajectory, which are represented by solid lines in Figure 6 has the interesting property of alternating energy absorption and energy regeneration phases between joints 1 and 2. The same figure shows that the amount of energy that can be regenerated by the fourth axis is negligible, given that the current regenerated during the braking phase by the linear unit is almost completely dissipated in the motor windings due to resistive losses.

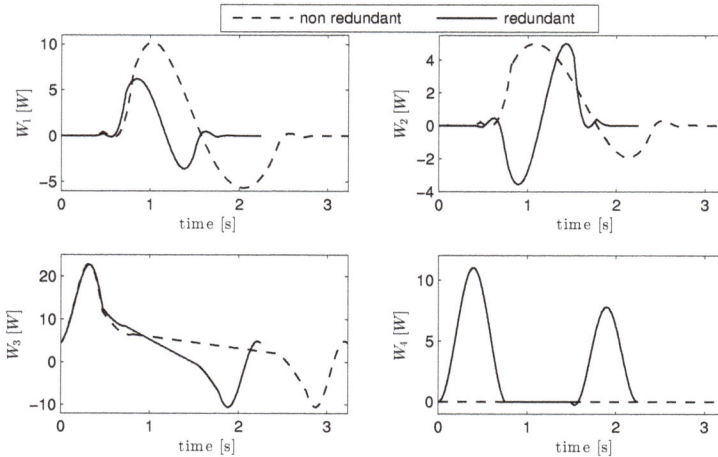

Figure 6. Motor electric power: Comparison between the energy-optimal trajectories with redundant and non-redundant configuration.

3.2. Trajectory Optimization Without Regeneration

The trajectory design procedure has been repeated by taking into consideration the very same task, with the only difference being that is has been assumed that the drive circuit of the SCARA robot do not support energy regeneration. The iteration of the design procedure for total execution times that range from 7 to 2 s, results in the two energy vs. time profiles shown in Figure 7. As far as the overall energy consumption is concerned, the addition of the fourth axis does, again, offers the possibility of reducing the energy consumption and achieving a consistent speed-up.

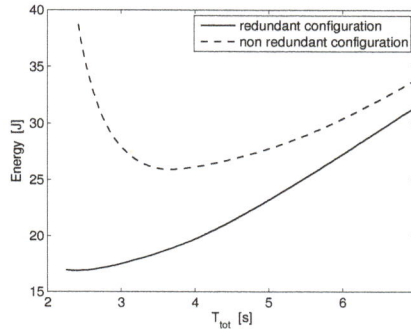

Figure 7. Energy consumption vs. total execution time: Comparison between the redundant and non-redundant configuration, without regenerative actuation.

The two energy-optimal solutions require 25.9 J and 16.9 J, with execution times equal to, respectively, 3.700 s and 2.442 s. The joint speed and acceleration profiles corresponding to the two optimal solutions are shown in Figures 8 and 9. Comparing them to Figures 3 and 4, which are generated under the full regeneration hypothesis, shows that the trajectories that lead to the energy optimality are very similar to each other. It can be inferred that, for the case under investigation, the energy regeneration has a minor effect on the trajectory design.

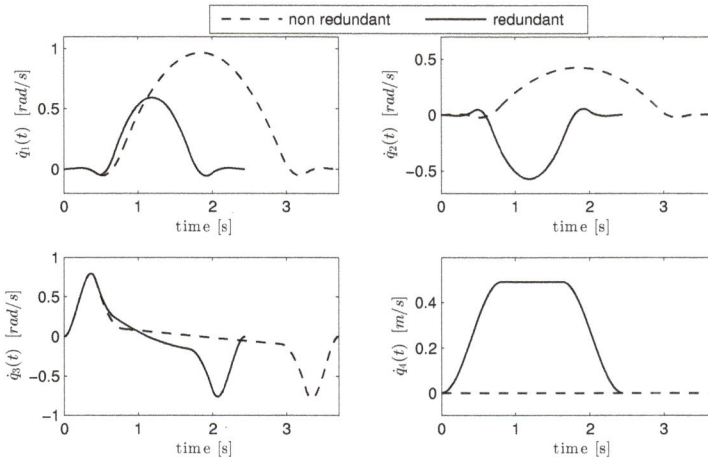

Figure 8. Joint speed: Comparison between the energy-optimal trajectories with redundant and non-redundant configuration, non-regenerative SCARA robot.

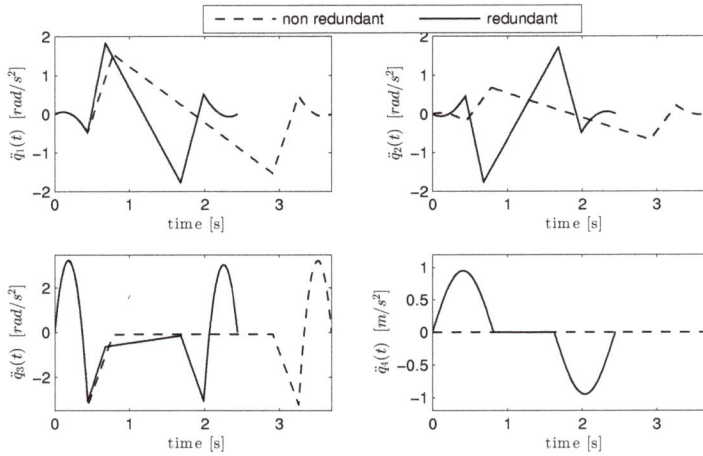

Figure 9. Joint acceleration: Comparison between the energy-optimal trajectories with redundant and non-redundant configuration, non-regenerative SCARA robot.

Figure 9 highlights that, accordingly to the results already presented in Figure 4, the timing of the motion of the linear unit that leads to the minimum energy consumption is the one that sets the acceleration and deceleration times equal to one third of the total execution time T.

Also the direct comparison between Figures 6 and 10, which show the power absorption with and without regeneration highlights the similarity between the two profiles. The absence of regenerated energy is indicated in Figure 10 by shading the areas underlined by negative values of the absorbed electric power: The energy associated to such contributions will be dissipated on a braking resistor and therefore will not be accounted for when evaluating Equation (5).

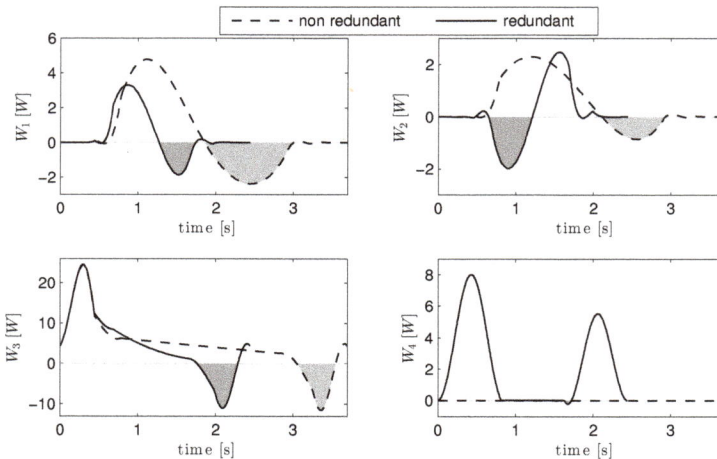

Figure 10. Motor electric power: Comparison between the energy-optimal trajectories with redundant and non-redundant configuration, non-regenerative SCARA robot. Dissipated energy is shown in gray.

The main difference between the optimal trajectories obtained with and without energy regeneration is that the latter requires a slightly longer execution time.

4. Conclusions

This works suggests a solution to the problem of designing optimal tasks for robotic cells by exploiting kinematic redundancy as a tool for improving the performance. The suggested method, applied to a robotic cell which comprises a SCARA robot and a linear unit used to move the workpiece, has been capable of producing motion profiles with a reduced energy consumption and a faster execution time in comparison to equivalent non-redundant configurations. Time speed-up and efficiency increase are achieved with and without the aid of regenerative braking. The trajectory optimization design is based on an inverse dynamic model of the robotic cell, which is used together with an analytical model of the energy consumption. The proposed method is of general application, since it can be applied to several other robotic architectures and to different motion profiles.

Author Contributions: Conceptualization, methodology, formal analysis, writing, editing: P.B. and D.R.; software, implementation, validation: P.B.

Funding: This research received no external funding.

Conflicts of Interest: The authors declare no conflict of interest.

References

1. Helm, D. The European framework for energy and climate policies. *Energy Policy* **2014**, *64*, 29–35. [CrossRef]
2. Tanaka, K. Review of policies and measures for energy efficiency in industry sector. *Energy policy* **2011**, *39*, 6532–6550. [CrossRef]
3. Carabin, G.; Wehrle, E.; Vidoni, R. A review on energy-saving optimization methods for robotic and automatic systems. *Robotics* **2017**, *6*, 39. [CrossRef]
4. Hansen, C.; Eggers, K.; Kotlarski, J.; Ortmaier, T. Comparative evaluation of energy storage application in multi-axis servo systems. In Proceedings of the 14th IFToMM world congress, Taipei, Taiwan, 25–30 October 2015; pp. 201–210.
5. Meike, D.; Ribickis, L. Energy efficient use of robotics in the automobile industry. In Proceedings of the 15th International Conference on Advanced Robotics (ICAR), Tallinn, Estonia, 20–23 June 2011; pp. 507–511.
6. Kim, Y.J. Design of low inertia manipulator with high stiffness and strength using tension amplifying mechanisms. In Proceedings of the 2015 IEEE/RSJ International Conference on Intelligent Robots and Systems (IROS), Hamburg, Germany, 28 September–2 October 2015; pp. 5850–5856.
7. Aziz, M.A.; Zhanibek, M.; Elsayed, A.S.; AbdulRazic, M.O.; Yahya, S.; Almurib, H.A.; Moghavvemi, M. Design and analysis of a proposed light weight three DOF planar industrial manipulator. In Proceedings of the IEEE Industry Applications Society Annual Meeting, Portland, OR, USA, 2–6 October 2016; pp. 1–7.
8. Li, Y.; Bone, G.M. Are parallel manipulators more energy efficient? In Proceedings of the 2001 IEEE International Symposium on Computational Intelligence in Robotics and Automation, Banff, AB, Canada, 29 July–1 August 2001; pp. 41–46.
9. Abe, A. An effective trajectory planning method for simultaneously suppressing residual vibration and energy consumption of flexible structures. *Case Stud. Mech. Syst. Sig. Process.* **2016**, *4*, 19–27. [CrossRef]
10. Brossog, M.; Kohl, J.; Merhof, J.; Spreng, S.; Franke, J. Energy consumption and dynamic behavior analysis of a six-axis industrial robot in an assembly system. *Procedia CIRP* **2014**, *23*, 131–136.
11. Recoskie, S.; Lanteigne, E.; Gueaieb, W. A High-Fidelity Energy Efficient Path Planner for Unmanned Airships. *Robotics* **2017**, *6*, 28. [CrossRef]
12. Alajlan, A.; Elleithy, K.; Almasri, M.; Sobh, T. An Optimal and Energy Efficient Multi-Sensor Collision-Free Path Planning Algorithm for a Mobile Robot in Dynamic Environments. *Robotics* **2017**, *6*, 7. [CrossRef]
13. Stuart, S. *DC Motors, Speed Controls, Servo Systems: An Engineering Handbook*, 3rd ed.; Pergamon Press: Oxford, UK, 1972.

14. Inoue, K.; Ogata, K.; Kato, T. An efficient power regeneration and drive method of an induction motor by means of an optimal torque derived by variational method. *IEEE Trans. Ind. Appl.* **2008**, *128*, 1098–1105. [CrossRef]

15. Hansen, C.; Öltjen, J.; Meike, D.; Ortmaier, T. Enhanced approach for energy-efficient trajectory generation of industrial robots. In Proceedings of the IEEE International Conference on Automation Science and Engineering (CASE), Seoul, Korea, 20–24 August 2012; pp. 1–7.

16. Hansen, C.; Kotlarski, J.; Ortmaier, T. Experimental validation of advanced minimum energy robot trajectory optimization. In Proceedings of the 16th International Conference on Advanced Robotics (ICAR), Montevideo, Uruguay, 25–29 November 2013; pp. 1–8.

17. Halevi, Y.; Carpanzano, E.; Montalbano, G. Minimum energy control of redundant linear manipulators. *J. Dyn. Syst. Meas. Contr.* **2014**, *136*, 051016. [CrossRef]

18. Ruiz, A.G.; Fontes, J.V.; da Silva, M.M. The influence of kinematic redundancies in the energy efficiency of planar parallel manipulators. In Proceedings of the ASME 2015 International Mechanical Engineering Congress and Exposition. American Society of Mechanical Engineers, Houston, TX, USA, November 13–19 2015; p. V04AT04A010.

19. Ruiz, A.G.; Santos, J.C.; Croes, J.; Desmet, W.; da Silva, M.M. On redundancy resolution and energy consumption of kinematically redundant planar parallel manipulators. *Robotica* **2018**, *36*, 809–821. [CrossRef]

20. Lee, G.; Sul, S.K.; Kim, J. Energy-saving method of parallel mechanism by redundant actuation. *Int. J. Precis. Eng. Manuf.-Green Technol.* **2015**, *2*, 345–351. [CrossRef]

21. Lee, J.; Lee, G.; Oh, Y. Energy-efficient robotic leg design using redundantly actuated parallel mechanism. In Proceedings of the IEEE International Conference on Advanced Intelligent Mechatronics (AIM), Munich, Germany, 3–7 July 2017; pp. 1203–1208.

22. Hansen, C.; Kotlarski, J.; Ortmaier, T. A concurrent optimization approach for energy efficient multiple axis positioning tasks. *J. Control Decis.* **2016**, *3*, 223–247. [CrossRef]

23. Sayed-Ahmed, A.; Wei, L.; Seibel, B. Industrial regenerative motor-drive systems. In Proceedings of the Twenty-Seventh Annual IEEE Applied Power Electronics Conference and Exposition (APEC), Orlando, FL, USA, 5–9 Febebruary 2012; pp. 1555–1561.

24. Richiedei, D.; Trevisani, A. Analytical computation of the energy-efficient optimal planning in rest-to-rest motion of constant inertia systems. *Mechatronics* **2016**, *39*, 147–159. [CrossRef]

25. Biagiotti, L.; Melchiorri, C. *Trajectory Panning for Automatic Machines and Robots*; Springer Science & Business Media: Berlin, Germany, 2008.

26. Ata, A.A. Optimal trajectory planning of manipulators: A review. *J. Eng. Sci. Technol.* **2007**, *2*, 32–54.

27. Cook, C.; Ho, C. The application of spline functions to trajectory generation for computer-controlled manipulators. In *Computing Techniques for Robots*. Springer: Berlin, Germany, 1984; pp. 101–110.

28. Boscariol, P.; Gasparetto, A.; Vidoni, R. Planning continuous-jerk trajectories for industrial manipulators. In Proceedings of the ASME 2012 11th Biennial Conference on Engineering Systems Design and Analysis. American Society of Mechanical Engineers, Nantes, France, 2–4 July 2012; pp. 127–136.

robotics

MDPI

Article

Cable Failure Operation Strategy for a Rehabilitation Cable-Driven Robot †

Giovanni Boschetti [1,*], **Giuseppe Carbone** [2] **and Chiara Passarini** [1]

[1] Department of Management and Engineering—DTG, University of Padova, 36100 Vicenza, Italy; chiara.passarini@phd.unipd.it

[2] Department of Mechanical, Energy and Management Engineering, University of Calabria, 87036 Rende, Italy; giuseppe.carbone@unical.it

* Correspondence: giovanni.boschetti@unipd.it; Tel.: +39-0444-99-8748

† This paper is an extended version our paper published in Boschetti, G.; Carbone, G.; Passarini, C. A Fail-Safe Operation Strategy for LAWEX (LARM Wire driven EXercising device). In the Proceedings of the 4th IFToMM Symposium on Mechanism Design for Robotics, Udine, Italy, 11–13 August 2018.

Received: 24 January 2019; Accepted: 2 March 2019; Published: 6 March 2019

Abstract: Cable-Driven Parallel Robots (CDPR) have attracted significant research interest for applications ranging from cable-suspended camera applications to rehabilitation and home assistance devices. Most of the intended applications of CDPR involve direct interaction with humans where safety is a key issue. Accordingly, this paper addresses the safety of CDPRs in proposing a strategy to minimize the consequences of cable failures. The proposed strategy consists of detecting a cable failure and avoiding any consequent motion of the end-effector. This is obtained by generating a wrench that is opposite to the direction of the ongoing motion so that the end-effector can reach a safe position. A general formulation is outlined as well as a specific case study referring to the LAWEX (LARM Wire-driven EXercising device), which has been designed within the AGEWELL project for limb rehabilitation. Real-time calculation is carried out for identifying feasible cable tensions, which generate a motion that provides the desired braking force. Simulations are carried out to prove the feasibility and effectiveness of the strategy outlined here in cases of cable failure.

Keywords: cable-driven parallel robots; fail-safe operation; exercising device

1. Introduction

Cable-driven parallel robots (referred to as CDPR in this paper) are parallel robots where the end-effector is linked to the base platform by replacing traditional rigid links with cables. This provides significant advantages, for example, in speed, acceleration, and workspace as compared with traditional robots. The abovementioned features make CDPRs an attractive alternative to traditional parallel robots. Cable-suspended cameras are widely known and even commercially implemented as applications of CDPR [1]. Moreover, many prototypes have been proposed in fields such as rehabilitation and home care [2]. Besides their numerous advantages, CDPRs introduce an additional constraint to motion planning and control, since cables can only exert positive tension forces on the end-effector. This characteristic makes trajectory planning and control tasks very complex. Moreover, CDPRs are more difficult to brake as compared with traditional robots with rigid links [3]. In fact, in traditional robots a hard stop of motors quickly leads to a full stop of the whole robot, while this cannot be applied to CDPRs since a hard stop of motors in CDPRs cannot prevent further motion of the flexible cables that allow motion in the robot end-effector. This makes it very difficult to create an emergency stop for CDPRs, while this feature would be very important from a safety point of view.

The failure of cables has hardly been addressed in the literature on cable-driven robots. An example can be found in [4], where the authors investigated the possibility of removing a cable while maintaining a static equilibrium pose. In particular, the authors of [4] proposed a method to identify the cables that are critical for ensuring the feasibility of a specific configuration. Then, the method finds out any unnecessary cable. In [5] and further in [6], Notash presented a possible classification of the feasible failure modes and introduced three methods to recover or compensate for a lost wrench of a redundant cable robot in static equilibrium conditions. In [7] the authors suggest using an optimal design procedure for minimizing the differences in cable tensions. This approach can also be applied in the case of a cable failure.

A completely different approach is presented in [8], where the authors propose a first method to control a moving end-effector after a cable failure. Such a strategy consists of generating a safe motion to a predetermined pose, while keeping the end-effector inside the workspace. The planning of such motion is based on pre-computed force limits and trajectories. A different approach was presented in [9], where the end-effector reaches the safe position by following a planned oscillatory trajectory. Such a trajectory still allows us to achieve a positive and bounded tension in each cable.

This paper improves on the approach proposed in [8] by proposing an efficient way to calculate and achieve a straight-line trajectory after failure, specifically for the case of the LAWEX robot [10,11]. This represents a first attempt to apply the Wrench Exertion Capability [12,13] as a tool to manage safety in cable-driven robots. It is important to note that this work is aiming at a preliminary feasibility study. Accordingly, only one type of cable failure has been considered, while other more complex cases will be considered as future work, with specific experimental validations with end users. The current work will preliminarily focus on the case of a single cable failure where the robot shape is preserved. In this case, even if the workspace keeps a shape similar to the original one, the configuration of the robot changes from an overconstrained to a suspended configuration. This change is very important, since a suspended CDPR must rely on gravity to maintain positive tension in all cables [14]. Moreover, when the robot configuration changes due to failure, its performance in terms of force exertion capability decreases unpredictably and the planned motion can become unfeasible. Hence, it is important to have a control tool that can be used in real time to estimate the feasible wrench and identify which cable tensions are needed to stop the end-effector in case of cable failure. Future works will investigate the possibility of applying the Wrench Exertion Capability to more complex CDPRs, as in [15].

The paper is organized as follows: Section 2 provides a description of the design and the key operation characteristics of LAWEX; Section 3 describes the proposed motion strategy in case of a cable failure; Section 4 provides a description of the main characteristics of LAWEX in terms of workspace and forces; Section 5 gives simulation results for the proposed motion strategy in case of one cable failure; finally, Section 6 gives the conclusions.

2. Characteristics of LAWEX Robot

In the last decade, a research team led by the second author at LARM has been investigating several cable-driven parallel robot architectures within the European co-funded project AGEWELL, currently underway at Technical University of Cluj-Napoca, Romania. Specifically, this project aims to investigate novel devices and design solutions for limb rehabilitation tasks, as reported for example in [10,11]. Within this frame a novel cable-driven design solution has been proposed and built as shown in Figure 1. The name of the prototype in Figure 1 is LAWEX (LARM Wire-Driven EXercising device). Its design has been based on preliminary studies that have been carried out on human patients to determine the desired limb motions, such as reported in [16–19]. The main innovative aspect of LAWEX is its open architecture, which can allow easy accessibility by users during treatment, even if they are sitting in a wheelchair. Aluminum profiles are used for a lightweight and stiff design, while the robot structure can be easily assembled, stored, and moved so that it can be suitable even for home use.

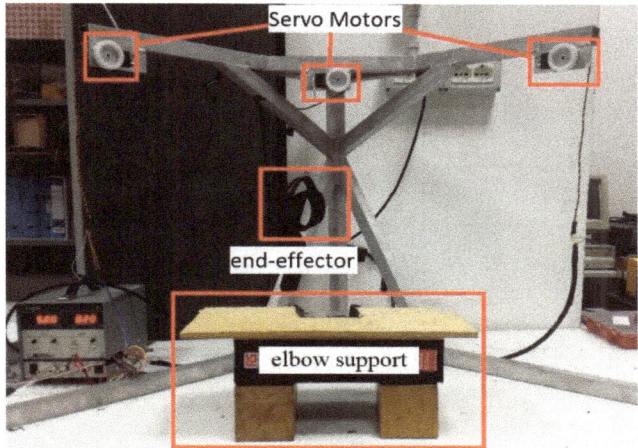

Figure 1. A built prototype of the LAWEX robot. (The main components are highlighted with red boxes.)

The LAWEX robot includes four servo motors and four cables; the cables are connected to the end-effector through a wristband as shown in Figure 1. Moreover, Figure 2 outlines an operation scheme of LAWEX. Namely, the human arm is attached to an interface/end-effector. This interface/end-effector is attached to a rigid frame by means of four cables. Each cable is connected to a servomotor that can change the cable length by using a winch. One of the servomotors is below the platform; the other three are attached to the upper part of the rigid frame. Cables are connected to the end-effector on the arm to be trained using a wristband. Further details on the end-effector are reported in [17]. This setup allows several different motions that are suitable for limb exercises such as for rehabilitation purposes.

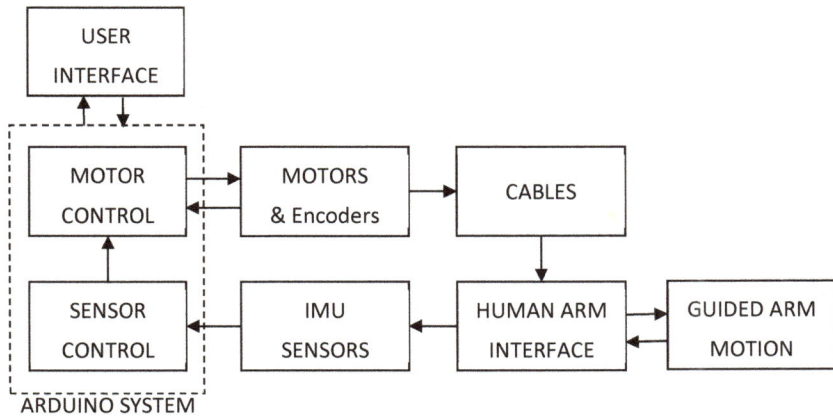

Figure 2. A scheme of the operation architecture of LAWEX.

A successful rehabilitation procedure relies on several key aspects. In particular, it is necessary to define and complete a reliable and safe motion based on a proper training protocol. Accordingly, several experiments have been carried out to set up the proper motion parameters such as the elbow support position and arm motion ranges. Some tested setup conditions are shown in Figure 1, where an elbow support keeps the elbow position fixed relative to LAWEX. This limits the arm motion to elbow flexion–extension.

2.1. Kinematics

The LAWEX frame structure has one servomotor in its bottom edge, where the fixed reference frame XYZ is also located. Three other servomotors are attached on the upper part of the robot frame structure, one at the top edge and the other two at the top extremities of the structure, as also shown in Figure 1. A kinematic scheme of LAWEX is shown in Figure 3, which gives the location of cables. In this figure the starting points of cables are indicated by using the position vector A_i ($i = 1, 2, 3, 4$), where each starting point refers to the attachment of a cable to its pulley that is attached to the corresponding servomotor. Accordingly, one can write:

$$A_1 = (-a_1; -a_2; h), \ A_2 = (0; 0; h), \ A_3 = (a_1; -a_2; h), \ A_4 = (0; 0; 0). \tag{1}$$

The pose of the end-effector EE in Figure 3 is defined by the cable connection points B_{ir}, ($i = 1, 2, 3, 4$), whose positions can be defined in reference to point H in the center of EE (this can also be defined as the mobile platform), such as:

$$B_{1r} = (-b_1; -b_2; 0), \ B_{2r} = (0; b_3; 0), \ B_{3r} = (b_1; -b_2; 0), \ B_{4r} = (0; 0; 0). \tag{2}$$

If one assumes a fixed orientation of the moving platform during the robot operation, the absolute position of point $H(x; y; z)$ can be easily obtained by means of a loop-closure equation for the i-th kinematic chain such as:

$$H + B_{ir} = A_i + E_i, \tag{3}$$

where E_i is the vector representing the i-th cable of the robot. Therefore, considering Equation (3), one can write:

$$E_i = H + B_{ir} - A_i. \tag{4}$$

In this way it is possible to solve the inverse kinematics problem of the cable robot by computing the vector modules on both sides as:

$$||E_i|| = ||H + B_{ir} - A_i||. \tag{5}$$

The relationships between the cable lengths and the position of the point H on the EE can be expressed as:

$$l_1 = \sqrt{(x + a_1 + b_1)^2 + (y + a_2 + b_2)^2 + (z - h)^2} \tag{6}$$

$$l_2 = \sqrt{x^2 + (y - b_3)^2 + (z - h)^2} \tag{7}$$

$$l_3 = \sqrt{(x - a_1 - b_1)^2 + (y + a_2 + b_2)^2 + (z - h)^2} \tag{8}$$

$$l_4 = \sqrt{x^2 + y^2 + z^2}. \tag{9}$$

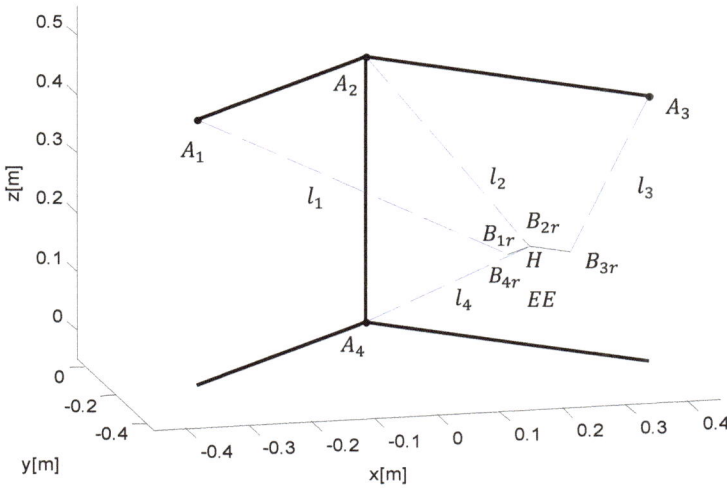

Figure 3. The kinematic scheme of the LAWEX architecture.

2.2. Dynamics

If one neglects the elasticity of the cables, the length of each cable can be considered as directly proportional to the angular displacement ϑ_i of the servomotors according to the relationship

$$l_i = l_0 - \vartheta r_0 - \frac{\pi r_1}{2} - d_0, \tag{10}$$

where l_0 is the total length of the cable, r_o is the driving pulley radio, r_1 is the secondary pulley radio, and d_0 is the distance between the centers of the pulleys. By isolating the term ϑ_i, it is possible to obtain the actuation vector of the manipulator for computing the corresponding motion angle, which allows the actuators to reach the desired position.

The dynamic model of this manipulator can be described by

$$w = \begin{bmatrix} f_x \\ f_y \\ f_z \end{bmatrix} = \begin{bmatrix} u_1 & u_2 & u_3 & u_4 & u_g \end{bmatrix} \begin{bmatrix} \tau_1 \\ \tau_2 \\ \tau_3 \\ \tau_4 \\ mg \end{bmatrix} = S\tau, \tag{11}$$

where w is the wrench vector that is represented by the three exerted forces f_i; S is the structure matrix that is made by the normal vectors u_i. Each vector u_i represents the direction that is connecting the end-effector with the i-th pulley; τ is the vector that contains the cable tensions τ_i. The abovementioned mg represents the gravity force, which can be considered as equivalent to a vector with constant direction and constant modulus.

2.3. Programming the LAWEX Operation

The programming of LAWEX has been developed according to Figure 4. In particular, Figure 4b shows a specifically developed user-friendly interface of LAWEX that runs in Java and can also be operated with Android smartphones. The user interface includes a button to perform the path planning. This is obtained by using the inverse kinematics of LAWEX, as discussed in the previous section and in [13]. The obtained results of the inverse kinematics are the cable lengths versus time, with which we are able to generate the desired path of the end-effector. The computed cable lengths are then converted into the angular positions of the servomotors. This is done by considering that the rotation of each servomotor winds or unwinds its cable about its own winch. Each winch consists of a cylindrical drum of known diameter. Accordingly, each servomotor rotation generates a known increase or decrease of the cable length. The computed vector of angular positions versus time is sent to an Arduino board, which drives all four servomotors.

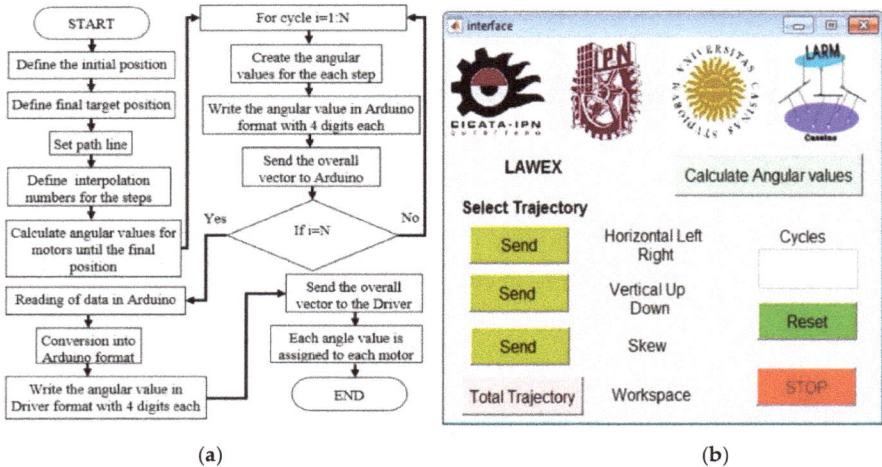

(a) (b)

Figure 4. The developed user interface of LAWEX: (a) flowchart of the software architecture; (b) the main interface screen.

The path planning can be generated in real time for short motions or off-line for predefined motion paths such as horizontal motions from left to right and vice versa, vertical motions up and down and vice versa, or a skew (diagonal) motion combining the previous two pre-defined paths. The latter achieves up–down motion followed by left–right motion. For each of these predefined paths it is possible to set a desired number of cycles N so that the path is cyclically repeated N times. The button "calculate angular values" starts the calculations for path planning and sends the output vector of angular positions to the Arduino board. One can set other trajectories through the "total trajectory" button by setting up a list of the point coordinates one wishes to reach versus time.

If the system has a failure, one can push the reset button and the end-effector returns to its initial configuration, which means it is at top up configuration and centered. One can also push the stop button in an emergency.

3. Motion Strategy after Failure

The main goal of the proposed strategy is planning and achieving a motion for reaching a safe position after a cable failure. In this application, such a strategy takes into account the Wrench Exertion Capability (WEC) of the rehabilitation robot and allows the motors to exert a braking force in the opposite direction of motion. To make sure that the strategy works, the end-effector should also be inside the feasible workspace after the cable failure. In the proposed case, this condition is certainly respected when cable 4 breaks: indeed, when another cable breaks, the workspace can degenerate into a bidimensional shape and the motion of the end-effector cannot be easily controlled. It is worth noting that, during a therapy rehabilitation task, cable 4 can reach tensions higher than the other cables. Therefore, it might be advisable to increase the number of cables to distribute the corresponding load on more cables in a redundant setup.

The force that can be exerted in a specific direction, before and after the cable failure, can be computed by means of a performance index called Wrench Exertion Capability (WEC). As mentioned above, this index computes the maximum force (or torque) that can be exerted in a given direction while keeping all the other components null. This index can be used to compute the maximum exertable braking force in the direction of motion, aimed at stopping the end-effector. It is important to note that a specific reference frame should be defined in order to compute the WEC index. Such a reference frame should have an axis that lies along the direction of motion d. Accordingly, the braking force is computed by searching for the cable tensions that guarantee the maximum braking force along the direction of motion d by referring to the axis of the abovementioned reference frame. The following linear programming problem allows for achieving the proper cable tension:

$$\text{minimize} : (f_d) = s_d \tau \tag{12}$$

$$s.t. : \begin{cases} S_o \tau = 0 \\ \tau_{min} \preccurlyeq \tau \preccurlyeq \tau_{max} \quad \forall \, \tau \neq \tau_f \\ \tau_f = 0 \end{cases} \tag{13}$$

where

- s_d is defined as the row of the Structure Matrix S_r that refers to the direction of motion. Let us define R as the rotation matrix that identifies the novel reference frame S_r by means of the following relation: $S_r = RS$;
- S_o is the matrix obtained by S_r by removing the row s_d;
- τ_{min} and τ_{max} are the vectors that define the range of the proper cable tensions;
- τ_f identifies the broken cable, whose tension is set to a null value.

From a practical point of view, following a cable failure, a wise choice can be to reduce the value of the maximum allowed cable tension in order to reduce the risk of another cable failure. Therefore, in this application we assume the maximum allowed cable tension after failure to be reduced to 70% of the maximum allowed cable tension before failure so that $\tau_{maxE} = 0.7 \times \tau_{max}$.

An important issue that is related to optimization problems is that they are iterative and time-consuming. For this reason, their application in a real-time environment is discouraged and sometimes not applicable. This aspect is thus carefully considered to keep the computational costs within a feasible range for real-time implementation. Accordingly, the proposed case study considers a computationally efficient approach for computing the WEC index and the maximum allowed braking force as soon as a cable failure is detected. The chosen method for computing the WEC index has already been proposed in [13]. It is based on a geometric representation of the forces that can be exerted. The main idea behind the algorithm is to consider the possible exertable wrenches as a polytope in the n-dimensional space (where n is the number of degrees of freedom). The n-dimensional space is represented so that the x-axis is oriented with the direction of interest d. The intersections between the x-axis and the skull of the wrench polytope represent the maximum and minimum exertable forces, respectively, in the direction of interest. The proposed algorithm also allows the corresponding tension configuration able to exert such forces. The corresponding tension vector is in the skull of the tension polytope in the m-dimensional case (where m is the number of cables). This algorithm allows for cutting 80% of the computational time compared with classic optimization-based algorithms.

4. Performance Analysis

This section reports a preliminary analysis of the performance (in terms of workspace and exertable force) of the cable robot before and after failure. The structure of the robot is described in Section 2. Considering the sizes of LAWEX, parameters a_1, a_2 are both set to 0.36 m; h is 0.46 m, while b_i can be neglected. For the performance analysis the origin of the fixed reference frame is attached to point A_2. Figure 5 represents the static equilibrium workspace before and after failure. Figure 5 is obtained by setting a minimum tension of 0.5 N and a maximum acceptable tension of 10 N as referring to the twisted iron cables of LAWEX. The mass of the end-effector is equal to 5 N (robot unload). On the other hand, the force exertion capability has been investigated by considering the directions of motion that the robot is supposed to have, i.e., left–right and up–down. The left–right motion in the considered reference frame results in a straight-line motion in the direction 45° in the plane x-y. The cyan rectangle in Figure 5 represents the plane for which the performance regarding the right–left motion has been investigated. Simulations suggest that the best performance for this motion can be reached when z = −h/2. Figure 6 shows the computed maximum exertable force in the right direction (a–c) and in the left direction (b–d). This has been computed before (a–b) and after (c–d) the failure of the cable with a color indicating the force magnitude. The color scale in [N] is reported on the right side of Figure 6. The shape of the polytope changes according to the position of the end-effector, since it is strictly related to the structure matrix **S**. It is important to note that only the tension polytope after failure is shown here, since it has a three-dimensional shape. Instead, the tension polytope before failure is a four-dimensional polytope that is not representable in a 3D plot.

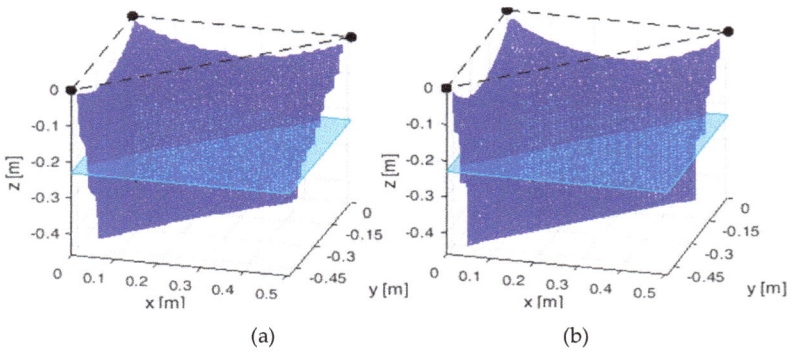

Figure 5. Workspace of LAWEX before (**a**) and after (**b**) failure.

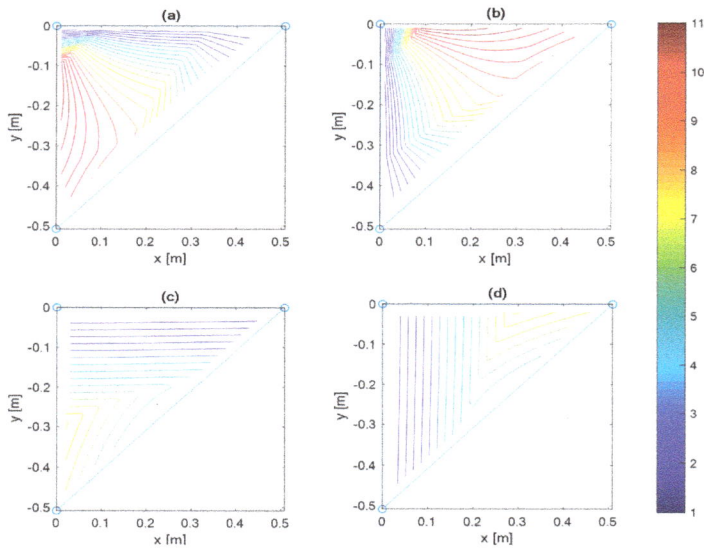

Figure 6. Computed maximum exertable force in the right direction (**a–c**) and in the left direction (**b–d**) before (**a,b**) and after (**c,d**) the failure of the cable.

5. Simulations

This section addresses the simulation of the behavior of LAWEX within its operation workspace. Discussion is also provided to clarify the main differences between the proposed approach and other previous approaches that are available in the literature, such as those reported in [5,6].

5.1. Simulation with Previous Approaches

Previous approaches are conceptually totally different from the proposed one as they aim to compensate for the lost wrench due to a cable failure by mostly considering the static equilibrium conditions.

A numerical simulation is reported to clarify the main features of the approaches in literature by referring to the case study of LAWEX. Namely, one can consider the LAWEX robot with the end-effector lying within its static feasible workspace. Without lack of generality, a feasible generic point P has been chosen with coordinates P (0.05, −0.2, 0.23). In this position the static equilibrium is guaranteed by the vector of cable tensions τ that allows one to achieve a null wrench w in Equation (11) so that:

$$w = \begin{bmatrix} u_1 & u_2 & u_3 & u_4 & u_g \end{bmatrix} \begin{bmatrix} \tau_1 \\ \tau_2 \\ \tau_3 \\ \tau_4 \\ mg \end{bmatrix} = S\tau = \begin{bmatrix} 0 \\ 0 \\ 0 \end{bmatrix}. \tag{14}$$

Since the robot is redundant, there are infinite tension vectors that satisfy Equation (14). In this case the solution with the minimum tension is chosen. Indeed, in this case one of the cables exerts the minimum allowed tension, 0.5 N, so that the vector of cable tensions τ is: $\tau = \{2.62, 2.94, 3.67, 0.5, mg\}$.

If one considers the failure of one cable, its tension becomes zero. In case of failure of the fourth cable, its tension became null ($\tau_4 = 0$). The existing methods in the literature, such as [5,6], aim at achieving a new set of cable tensions that satisfies Equation (4) under the condition that one cable is no longer able to exert any force. Accordingly, one can compute the wrench vector w as follows:

$$w = \begin{bmatrix} u_1 & u_2 & u_3 & u_4 & u_g \end{bmatrix} \begin{bmatrix} \tau_1 \\ \tau_2 \\ \tau_3 \\ 0 \\ mg \end{bmatrix} = S\tau = \begin{bmatrix} 0 \\ 0 \\ 0 \end{bmatrix}. \tag{15}$$

In this case, the robot is no longer redundant, since one cable is considered broken. Accordingly, there is only one feasible tension vector that satisfies Equation (15) and vector τ can be computed as $\tau = \{2.29, 2.99, 3.19, 0, mg\}$. This new tension vector allows for maintaining the static equilibrium of the end-effector after a cable failure.

5.2. Simulation of the Proposed Motion Strategy

The proposed approach differs totally from other existing methods, since it deals with a cable failure by planning a motion strategy under dynamic conditions, instead of addressing the static case as in existing methods. A numerical simulation is reported here to clarify the main features of the proposed novel approach by referring to the case study of LAWEX.

An algorithm has been written in Matlab-Simulink in order to prove its effectiveness. In a first test, a periodic motion from left to right has been considered. Such a motion has been defined by considering the center of the end effector starting at coordinates $X_i = [0.05, −0.35, −0.23]m$ and stopping its motion at coordinates $X_f = [0.35, −0.05, −0.23]m$; the period of the full motion (roundtrip) requires about 2 s equally distributed in the forward motion (lasting one second) and the backward motion (lasting 1 s). We have considered a failure of cable 4 occurring at $t = 0.65$ s. In Figure 7 the full motion performed by the end-effector is shown by a blue line, while the instant of failure is depicted in red.

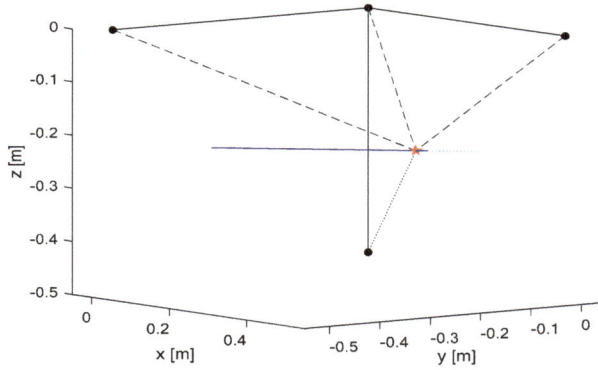

Figure 7. The computed end-effector trajectory for the proposed simulation test.

Dashed lines in Figure 7 refer to cables that are still active after the failure of another cable. The damaged or broken cable has been highlighted with a dotted line. The absolute velocity of the whole path has been plotted in Figure 8 by showing the motion before the cable failure as phase 1 (from 0 to 0.65 s); the motion after cable failure as phase 2 (from 0.65 to 0.7 s); the reaching of final configuration as phase 3 (from 0.7 to 1 s). The tensions of the cables are depicted in Figure 9. The time axis lasts 1 s in order to highlight the three different phases. During the first phase the end-effector is moving along the planned trajectory; in the second phase, after the cable failure, the proposed algorithm computes the cable tensions that are needed for exerting the braking force that is needed to stop the end-effector. The last phase begins when the velocity of the end-effector is close to zero. In this phase the cable tension reaches a static equilibrium configuration in the actual position.

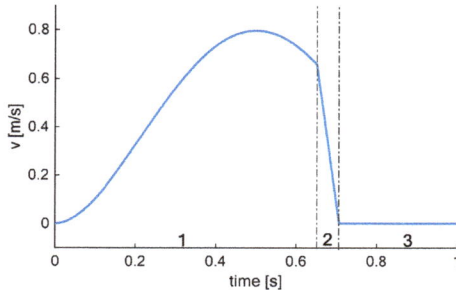

Figure 8. The end-effector velocity during the simulation (phase 1: motion before the cable failure; phase 2: motion after cable failure; phase 3: reaching of final position).

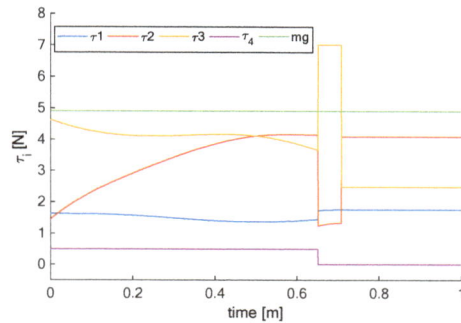

Figure 9. Cable tensions during the simulation (phase 1: motion before the cable failure; phase 2: motion after cable failure; phase 3: reaching of final position).

6. Conclusions

This paper addresses the safe operation of cable-driven parallel manipulators by taking into account the effect of a cable failure. Specifically, we propose a strategy to deal with one cable failure. This strategy consists of a real-time computation that, after detecting a cable failure, generates a cable tension that can achieve quick braking of the end-effector in a safe pose. A computation algorithm has been discussed and simulation results have been analyzed referring to the real case study of a LAWEX robot. This has allowed us to confirm the engineering feasibility and effectiveness of the proposed strategy when a cable failure occurs.

Author Contributions: Conceptualization, G.B., G.C. and C.P.; Investigation G.B., G.C. and C.P.; Methodology, G.B., G.C. and C.P.; G.B., G.C. and C.P. wrote the paper.

Funding: This work was supported by the Project ID 37_215, MySMIS code 103,415 "Innovative approaches regarding the rehabilitation and assistive robotics for healthy ageing" co-financed by the European Regional Development Fund through the Competitiveness Operational Program 2014-2020, Priority Axis 1, Action 1.1.4, through the financing contract 20/01.09.2016, between the Technical University of Cluj-Napoca and ANCSI as Intermediary Organism in the name and for the Ministry of European Funds.

Conflicts of Interest: The authors declare no conflict of interest.

References

1. Cone, L.L. Skycam-an aerial robotic camera system. *Byte* **1985**, *10*, 122.
2. Rosati, G.; Gallina, P.; Rossi, A.; Masiero, S. Wire-based robots for upper-limb rehabilitation. *Int. J. Assist. Robot. Mechatron.* **2006**, *7*, 3–10.
3. Boschetti, G. A picking strategy for circular conveyor tracking. *J. Intell. Robot. Syst.* **2016**, *81*, 241–255. [CrossRef]
4. Roberts, R.G.; Graham, T.; Lippitt, T. On the inverse kinematics, statics, and fault tolerance of cable-suspended robots. *J. Field Robot.* **1998**, *15*, 581–597. [CrossRef]
5. Notash, L. Failure recovery for wrench capability of wire-actuated parallel manipulators. *Robotica* **2012**, *30*, 941–950. [CrossRef]
6. Notash, L. Wrench recovery for wire-actuated parallel manipulators. *Proc. Romansy 19 Robot Des. Dyn. Control* **2013**, 201–208. [CrossRef]
7. Ghaffar, A.; Hassan, M. Failure analysis of cable based parallel manipulators. *Appl. Mech. Mater.* **2014**, *736*, 203–210. [CrossRef]
8. Boschetti, G.; Passarini, C.; Trevisani, A. A recovery strategy for cable driven robots in case of cable failure. *Int. J. Mech. Control* **2017**, *18*, 41–48.
9. Passarini, C.; Zanotto, D.; Boschetti, G. Dynamic Trajectory Planning for Failure Recovery in Cable-Suspended Camera Systems. *J. Mech. Robot.* **2019**. [CrossRef]

10. Carbone, G.; Gherman, B.; Ulinici, I.; Vaida, C.; Pisla, D. Design Issues for an Inherently Safe Robotic Rehabilitation Device. In *International Conference on Robotics in Alpe-Adria-Danube Region*; Spinger: Dorchect, The Netherlands, 2017; pp. 967–974.
11. Carbone, G.; Arostegui Cavero, C.; Ceccarelli, M.; Altuzarra, O. A study of Feasibility for a Limb Exercising Device. *Mech. Mach. Sci.* **2017**, *47*, 11–21.
12. Boschetti, G.; Trevisani, A. Cable robot performance evaluation by wrench exertion capability. *Robotics* **2018**, *7*, 15. [CrossRef]
13. Boschetti, G.; Passarini, C.; Trevisani, A.; Zanotto, D. A Fast Algorithm for Wrench Exertion Capability Computation. In *Cable-Driven Parallel Robots*; Springer: Cham, Switzerland, 2018; pp. 292–303.
14. Trevisani, A. Planning of dynamically feasible trajectories for translational, planar, and underconstrained cable-driven robots. *J. Syst. Sci. Complex.* **2018**, *26*, 695–717. [CrossRef]
15. Scalera, L.; Gallina, P.; Seriani, S.; Gasparetto, A. Cable-Based Robotic Crane (CBRC): Design and Implementation of Overhead Traveling Cranes Based on Variable Radius Drums. *IEEE Trans. Robot.* **2018**, *34*, 474–485. [CrossRef]
16. Major, K.A.; Major, Z.Z.; Carbone, G.; Pisla, A.; Vaida, C.; Gherman, B.; Pisla, D.L. Ranges of Motion as Basis for robot-assisted post-stroke rehabilitation. *Hum. Vet. Med.* **2016**, *8*, 192–196.
17. Carbone, G.; Ceccarelli, M.; Rodríguez León, J.F.; Lazar, V.A.; Cafolla, D.; Vaida, C.; Pisla, D. Experimental Characterization of Assisted Human Arm Exercises. In Proceedings of the 2018 IEEE International Conference on Automation, Quality and Testing, Robotics (AQTR), Cluj-Napoca, Romania, 24–26 May 2018.
18. Rodríguez León, J.F.; Carbone, G.; Cafolla, D.; Russo, M.; Ceccarelli, M.; Castillo Castañeda, E. Experiences and Design of a Cable-Driven Assisting Device for Arm Motion. *Cism Int. Cent. Mech. Sci. Courses Lect.* **2019**, *584*, 94–101.
19. Laribi, M.A.; Carbone, G.; Zeghloul, S. Optimal design of cable driven robot for rehabilitation with prescribed workspace. *Mech. Mach. Sci.* **2019**, *65*, 273–282.

robotics

MDPI

Article

V2SOM: A Novel Safety Mechanism Dedicated to a Cobot's Rotary Joints [†]

Younsse Ayoubi, Med Amine Laribi *, Said Zeghloul and Marc Arsicault

Department of GMSC, Pprime Institute CNRS, ENSMA, University of Poitiers, UPR 3346, 86073 Poitiers, France; younsse.ayoubi@univ-poitiers.fr (Y.A.); said.zeghloul@univ-poitiers.fr (S.Z.); marc.arsicault@univ-poitiers.fr (M.A.)

* Correspondence: med.amine.laribi@univ-poitiers.fr; Tel.: +33-(0)5 49 49 65 52

† This paper is an extended version of our paper published in Ayoubi, Y.; Laribi, M.A.; Zeghloul, S.; Arsicault, M. Design of V2SOM: The Safety Mechanism for Cobot's Rotary Joints. In Proceedings of the IFToMM Symposium on Mechanism Design for Robotics, Udine, Italy, 11–13 August 2018.

Received: 11 January 2019; Accepted: 25 February 2019; Published: 6 March 2019

Abstract: Unlike "classical" industrial robots, collaborative robots, known as cobots, implement a compliant behavior. Cobots ensure a safe force control in a physical interaction scenario within unknown environments. In this paper, we propose to make serial robots intrinsically compliant to guarantee safe physical human–robot interaction (pHRI), via our novel designed device called V2SOM, which stands for Variable Stiffness Safety-Oriented Mechanism. As its name indicates, V2SOM aims at making physical human–robot interaction safe, thanks to its two basic functioning modes—high stiffness mode and low stiffness mode. The first mode is employed for normal operational routines. In contrast, the low stiffness mode is suitable for the safe absorption of any potential blunt shock with a human. The transition between the two modes is continuous to maintain a good control of the V2SOM-based cobot in the case of a fast collision. V2SOM presents a high inertia decoupling capacity which is a necessary condition for safe pHRI without compromising the robot's dynamic performances. Two safety criteria of pHRI were considered for performance evaluations, namely, the impact force (ImpF) criterion and the head injury criterion (HIC) for, respectively, the external and internal damage evaluation during blunt shocks.

Keywords: cobot; V2SOM; safety mechanism; safe physical human–robot interaction; pHRI; variable stiffness actuator; VSA; collaborative robots

1. Introduction

Robotics was introduced into industry at the beginning of the 1960s. Several industries (e.g., automobile, military and manufacture) improved their productivity rates thanks to the use of robots, taking advantage of their capabilities to execute repetitive tasks much faster than humans. Those classical industrial robots generally executed the production tasks in highly secured cells, out of the reach of human operators. Nevertheless, other tasks cannot be easily automated, and human execution is therefore required, such as complex tasks or the manipulation of heavy loads. The use of collaborative robots, known as cobots, emerges as a solution to improve the execution of those tasks where a human is required. Unlike classical industrial robots, usually isolated to avoid physical contact with humans, cobots actually coexist with them in a shared common workspace and cooperate with them to accomplish the desired tasks. While a robot can magnify human capabilities, such as their force, speed, or precision, humans can bring a global knowledge and their experience to jointly execute the tasks [1]. With the fourth industrial revolution, the number of cobots has increased [2] and they are being used more and more to assist well-experienced humans.

Safety is the most important issue to solve before establishing collaborative tasks between humans and robots, where a high risk of collisions between them is evident and may result in damage to humans. In this context, research efforts are focused on the design of solutions to reduce the energy transferred by the robot in the case of collision, decreasing the risk of injury for the human [3]. In this regard, some basic solutions have been proposed. For instance, Park et al. introduce the use of a viscoelastic covering on the robot's body to reduce the impact forces [4]. Fritzsche et al. propose monitoring the contact forces by providing the robot's body with a tactile sensor used as an artificial skin [5]. Furthermore, several control approaches have been proposed to provide the robot with a compliant behavior while it executes a task. These compliant control strategies typically make it possible to assign a dynamic relationship between the robot and the environment, enabling the interaction behavior to be controlled by properly selecting the dynamic parameters. The compliant behavior can be either implemented in the robot end-effector or in the joints, for the Cartesian or joint space cases, respectively. A complete survey of the different collaborative control schemes can be found in [6]. On the other hand, mechanical solutions have also been proposed to provide an intrinsic compliance to the robot. Among these compliant mechanisms, variable stiffness actuators (VSAs) allow the introduction of an intrinsic compliance to the robot joints [7]. These mechanisms are capable of providing adjustable stiffness to the joints, which can be adjusted according to the needs.

Overall, two main approaches are well-respected for the human safety versus robot dynamics trade-off. These approaches are summarized under active impedance control and passive compliance (PC). The first approach suffers from a low latency in the case of blunt HR collision that reaches up to 200 ms [8,9], which may endanger human safety. In contrast, passive compliance presents a robust instantaneous response to uncontrolled HR shocks. In general, what makes robots intrinsically dangerous is the combination of high velocities and massive mobile inertia [10]. This latter aspect is a key feature in making cobots behave safely without limiting the desired dynamic performances, that is, by decoupling the cobot's colliding part inertia from the heavy rotor side inertia via passively compliant joints. In this respect, the earliest works yielded the series elastic actuator and the series parallel elastic actuator [11,12], where the stiffness is constant. As this behavior cannot cope with a cobot's load variation and its dynamics, Zinn proposed the concept DM^2 in [13] that improves the control via the double actuation system. Subsequently, the concept of variable stiffness actuator (VSA) gained more attention [8,14–17] from the robotics community. The VSA acts upon a wide range of a cobot's load by adapting its apparent stiffness. Note that every VSA is different, for example, in terms of its stiffness profile or working principal. A well-detailed study of a VSA's design goal is presented in [8]. There are some examples of systems implementing VAS control with a sensor-based approach, which usually leads to a more complex mechanical structure, and a sensorless approach, as commonly used in the position/stiffness control [18,19]. This study's proposed approach, leading to the prototype V2SOM [20], presents the following novelties with respect to the literature:

- The stiffness behavior, in the vicinity of zero deflection, is smoothened via a cam-follower mechanism.
- The stiffness sharply sinks to maintain, theoretically as shown in Figure 5, a constant torque threshold in the case of collision.
- The torque threshold, T_{max}, is tunable according to load variation.

The focus of this paper is V2SOM's design which is dedicated to a cobot's rotary joints. In Section 2, the mechanism's working principle is presented by emphasizing its design concept and the two functional blocks: stiffness generation block and stiffness adjusting block. The design methodology applied to obtain the first V2SOM prototype is presented in Section 3. The theoretical as well as experimental characteristics of the first V2SOM prototype are addressed at the end of that section. Section 4 presents a comparative study between V2SOM and a constant stiffness (CS) profile, comparing the choice of both HIC and ImpF criteria via simulation. Section 5 summarizes some notable outcomes and perspectives of the present study.

2. V2SOM's Working Principle

The design concept of the variable stiffness mechanism (VSM) aims to make load-adjustable compliant robots by implementing VSM in series with the actuation system, as depicted in Figure 1. However, a VSM can simply be described as a tunable spring with a basic nonlinear stiffness profile.

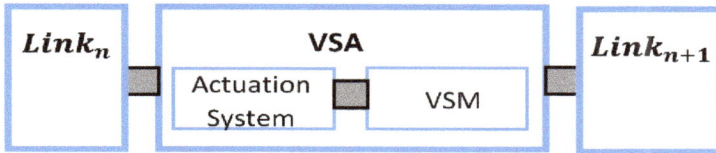

Figure 1. Variable stiffness actuator (VSA) scheme, including the actuation system coupled with a variable stiffness mechanism (VSM).

2.1. Architecture Description

V2SOM contains two functional blocks, as depicted in Figure 2, namely, a nonlinear stiffness generator block (SGB) and a stiffness adjusting block (SAB). To simplify the understanding of V2SOM, we presented a kinematic scheme in Figure 2b with a semi-view that is symmetrical to the rotation axis **L1**. The SGB is based on a cam-follower mechanism where the cam's rotation γ about **L5** axis, between $-90°$ and $90°$, induces the translation of its follower according to the slider **L6**. Then, the follower extends its attached spring. At this level, a deflection angle γ corresponds to a torque value T_γ exerted on the cam. The wide range of this elastic deflection must be reduced to a lower range of $-20° \leq \theta \leq 20°$, as it is widely considered in most VSAs [21,22]. To this end, the SAB acts as a reducer by using a gear ring system. Furthermore, the SAB serves as a variable reducer due to the linear actuator M that controls the distance a while driving the gearing in a lever-like configuration. The reduction ratio of the SAB is continuously tunable allowing V2SOM to cope with the external load T_θ, where the link side makes a deflection angle θ relative to the actuator side.

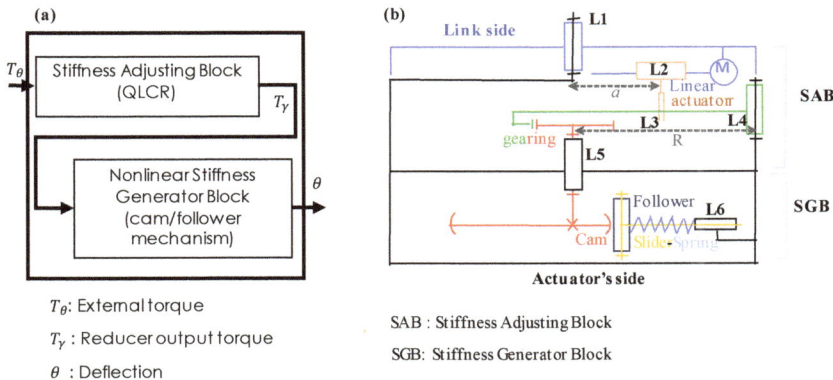

Figure 2. Architecture description: (**a**) Block representation of the V2SOM (**b**) Kinematic scheme in semi-view of the V2SOM.

The V2SOM CAD model as well as its first prototype are presented in Figure 3. More details are given in the V2SOM patent [20] and [15,16]. The V2SOM blocks as shown in Figure 3, with CAD models and corresponding prototypes, are connected rigidly to fulfil each step of a dedicated task:

- The SGB is characterized by the curve of the torque T_γ vs. the deflection angle γ. This curve is obtained through the cam profile, the followers. and other design parameters. The basic torque curve leading to the torque characteristic of the V2SOM is depicted in Figure 4a. This basic curve is elaborated with a torque threshold equal to $T_{max} = 2.05$ Nm, for the present prototype.

- The SAB is considered as a quasi-linear continuous reducer (QLCR) and defined by its ratio expression given in Figure 4b. The ratio is a function of the nonlinear (NL) factor, deflection angle γ, and reducer's tuning parameter a. The NL factor is linked to the SAB's internal parameters and can be approximated with a constant when the deflection θ range is between $-20°$ and $20°$; this issue will be discussed in the next section.

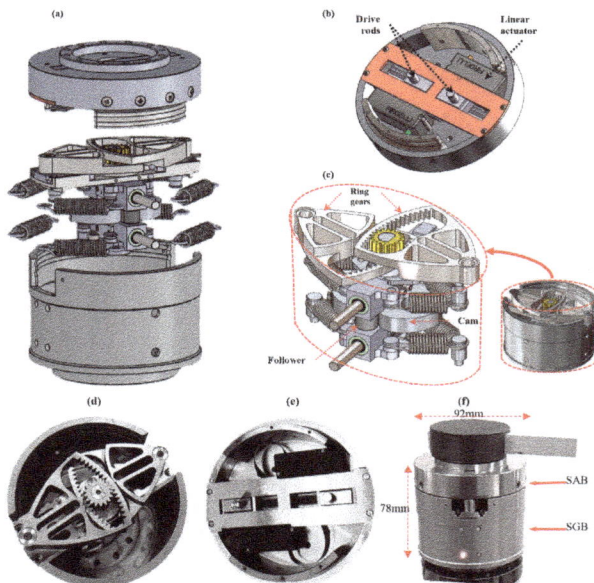

Figure 3. V2SOM: (**a**) 3D model, (**b**) CAD model of upper block, (**c**) CAD model of lower block, (**d**) prototype of upper block, (**e**) prototype of lower block, (**f**) first prototype.

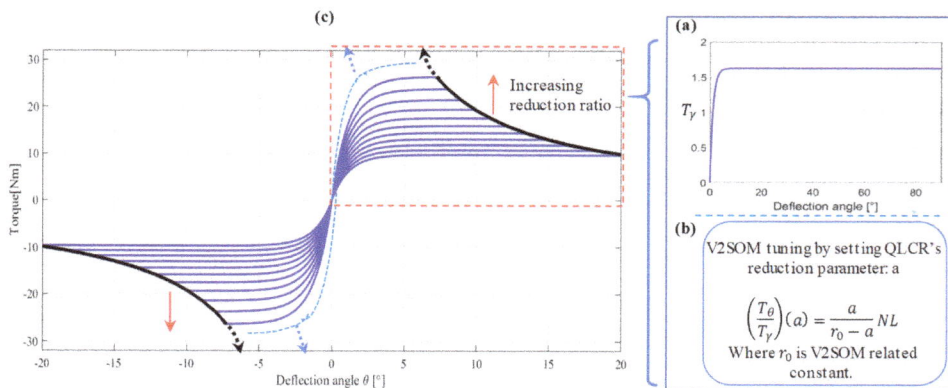

Figure 4. (**a**) Example of V2SOM basic torque curve with (**b**) QLCR. (**c**) Illustration of the V2SOM torque characteristic with seven QLCR settings.

Figure 4c shows an illustration of the V2SOM characteristic resulting from Figure 4a with seven increasing reduction ratio settings (seven values of torque tuning). Because of the QLCR behavior of the SAB, the curves in Figure 4c follow the profile of the basic torque curve given in Figure 4b which will be detailed in Section 2.2.

In general, V2SOM has two working modes between which a transition smoothly takes place in the case of blunt shock as illustrated in Figure 5. The high stiffness mode (I) is defined within the deflection range $[0, \theta_1]$ and the torque range $[0, T_1]$. The T_1 value defines the normal working conditions of the torque. Exceeding this torque value means that the shock absorbing mode (II) is triggered, characterized with low stiffness thus leading to the torque threshold T_{max}.

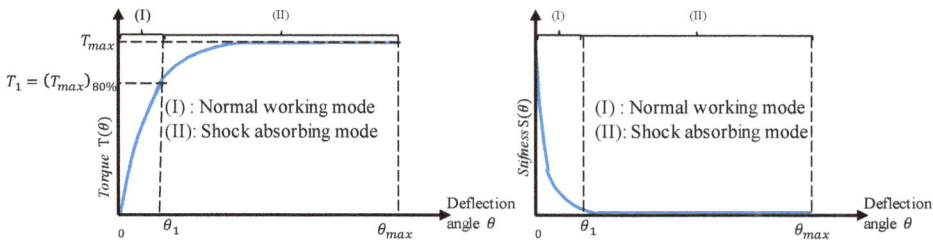

Figure 5. V2SOM working modes. (**Left**) Torque curve, (**Right**) stiffness curve.

2.2. Stiffness Generator Block (SGB)

In order to explain the working principle of the stiffness generator block, its corresponding simplified sketches are shown in Figure 6. A relaxed spring configuration with the cam-follower is presented in Figure 6a. In this particular configuration, the supported torque around the rotation axis is equal to zero, $T_\gamma = 0$. The increase of the torque $T_\gamma \nearrow$ leads to an elastic deflection angle as well as a linear motion of the sliders. The rotation of the cam, as depicted in Figure 6b, compresses the springs, allowing the followers to keep contact with the cam profile. Contact surface analysis allowed us identify the interaction forces and then establish the static equilibrium conditions. A graphic representation of a contact surface between cam and follower is shown in Figure 6c,d. Below, geometric parameters are listed in addition to the corresponding static equilibrium force equations.

- ρ: Distance among the rotation center, point O, and contact point of the cam-follower.
- r_1: Follower's radius.
- r: Distance of the follower's center, point O, to the cam's rotation center. The rest value is r_0.
- F_r : Resultant force at the cam-follower contact.
- F_c : Component of F_r in charge of deflection torque T_γ. The relations can be written as follows:

$$\begin{cases} T_\gamma = 2\,\rho F_c \\ F_c = \cos(\beta - \alpha)F_r \end{cases} \tag{1}$$

Notice that the components of F_r of each follower, according to the axis containing the rotation center, are cancelling each other.

- F_f: Friction force at the slider supporting the follower.
- F_k: Compression force of the spring.
- R: Force applied on the slider perpendicular to its axis.

The following equations can be deduced from the static equilibrium condition, giving the relation between forces F_f, F_k, R and F_r:

$$\begin{cases} R = \cos \alpha F_r \\ F_k + F_f = \sin \alpha F_r \end{cases} \tag{2}$$

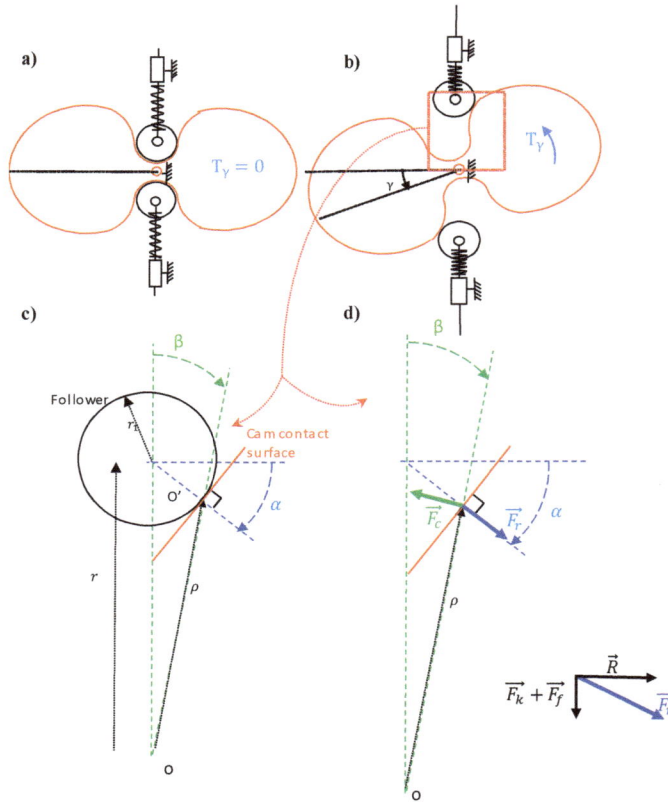

Figure 6. Stiffness generator block: (**a**) at rest $\gamma = 0$ (T_γ =0), (**b**) at deflection $\gamma \neq 0$ (T_γ), (**c**) cam-follower contact surface geometric parameters, (**d**) cam-follower contact forces.

Considering a Coulomb-type friction at the slider joint and a proportional expression to F_r and μ, the force and friction coefficient are expressed, respectively, as follows:

$$F_f = \mu R = \mu \cos \alpha F_r. \tag{3}$$

The following relations between the geometric parameters, useful for the equation rearrangement, can be obtained from Figure 6c:

$$\begin{cases} r_1 \cos \alpha = \rho \sin \beta \\ r - r_1 \sin \alpha = \rho \cos \beta \end{cases}. \tag{4}$$

Equation (4) is rearranged as follows:

$$\begin{cases} \beta = a \tan\left(\frac{r_1 \cos \alpha}{r - r_1 \sin \alpha}\right) \\ \rho = \frac{r_1 \cos \alpha}{\sin \beta} \end{cases}. \tag{5}$$

A rotation of the cam from equilibrium position, $0 \to \gamma$ yields to follower motion defined by a translation of the slider from initial position, $r_0 \to r$. The principle of virtual work applied to the SGB can be written as follows:

$$\int_0^\gamma T_\gamma(x)\, dx = 2 \int_{r_0}^r F_k(y) + F_f(z)\, dy. \tag{6}$$

Then, Equation (6) is simplified as follows:

$$\int_0^\gamma T_\gamma(x)\,dx = 2\int_{r_0}^r F_k(y)\left(1 + \frac{\mu\cos\alpha}{\sin\alpha - \mu\cos\alpha}\right)dy. \tag{7}$$

By substituting $x = \gamma(y)$ in Equation (7), we obtain:

$$\int_{r_0}^r \dot\gamma\, T_\gamma(\gamma(y))\,dy = 2\int_{r_0}^r F_k(y)\left(1 + \frac{\mu\cos\alpha}{\sin\alpha - \mu\cos\alpha}\right)dy, \tag{8}$$

where $\dot\gamma = \frac{\partial\gamma}{\partial r}$, $\gamma(r_0) = 0$ and $\gamma(a) = \gamma$.

As Equation (8) is valid for any $r \geq r_0$, one may write the following:

$$\dot\gamma T_\gamma(\gamma(r)) = 2F_k(r)\left(1 + \frac{\mu\cos\alpha}{\sin\alpha - \mu\cos\alpha}\right). \tag{9}$$

Combining Equations (1)–(3), (5), and (9) results in the following:

$$\begin{cases} \tan\alpha = r\dot\gamma \\ \dot\gamma = \frac{2F_k(r)}{T_\gamma(\gamma)} + \frac{\mu}{r} \end{cases} \quad \text{where } \gamma \neq 0. \tag{10}$$

The first step to find the cam's profile, that is, the set of points defined as the pair (β,ρ), is to solve the second differential Equation of (10). Then, the pairs (β,ρ) are found using Equation (5).

In this regard, the following numerical scheme is adopted:

$$\dot\gamma(r_i) = \frac{\gamma(r_{i+1}) - \gamma(r_i)}{r_{i+1} - r_i}. \tag{11}$$

Substituting (11) in the second Equation of (10), leads to:

$$\gamma_{i+1} = \gamma_i + (r_{i+1} - r_i)\left(\frac{2F_k(r_i)}{T_\gamma(\gamma_i)} + \frac{\mu}{r_i}\right), \tag{12}$$

where $\gamma_{i+1} = \gamma(r_{i+1})$, $\gamma_i = \gamma(r_i)$ and $\gamma_0 = \gamma(r_0) = 0$.

For the sake of simplicity, we presented a combination of four extension springs (see Figure 3b) as a single compression spring in Figure 6. Thus, the springs in Figure 6 have the following characteristic:

$$F_k = 4r\mathrm{K}\left(1 - \frac{l_0}{\sqrt{r^2 + b^2}}\right), \tag{13}$$

where K and l_0 are the extension springs' stiffness and initial length, respectively, and b is a constant, depicted in Figure 7, related to clash constraints.

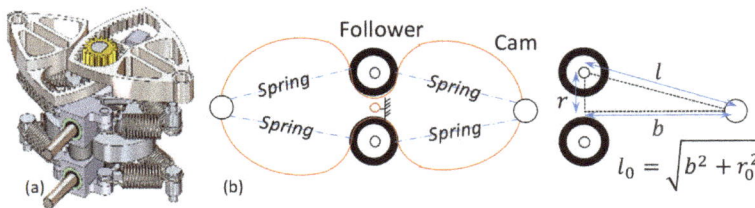

Figure 7. Spring setup in the lower block: (a) CAD model, (b) simple sketch and parameters.

It is now clear that the cam's profile depends on the characteristic of deflection torque T_γ vs. deflection angle γ. The SGB of V2SOM has the torque characteristic shown in Figure 8, where the two functional modes are shown. The equation of the curve in Figure 8 is given by the following:

$$T_\gamma(\gamma) = T_{max}\left(1 - e^{-s\gamma}\right), \tag{14}$$

where T_{max} and s are two constants of the designer's choice.

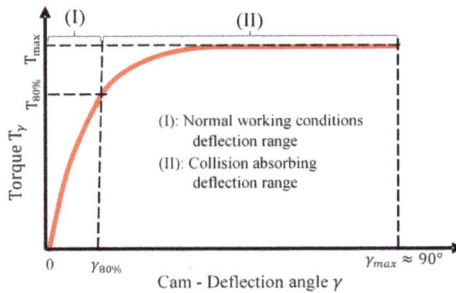

Figure 8. Torque vs. deflection characteristic of stiffness generating block and V2SOM's two functional modes.

The first mode, which we call normal operational condition mode (i.e., no collision takes place), is defined within the deflection range (I) of Figure 8. The range (I) is characterized by a torque value $T_{80\%}$ and a user chosen value, that represents 80% of T_{max}. Accordingly, the range (II) represents the collision mode where the deflection torque at the SGB exceeds the threshold of $T_{80\%}$.

2.3. Stiffness Adjusting Block (SAB)

The interval delimitating the deflection angle of the SGB is within $[-90°; 90°]$ which is larger then numerous cobot applications (e.g., $[-15°; 15°]$ for humanoid arm application [11]). The tuning of the V2SOM's stiffness in a continuous manner is supported by the stiffness adjusting block (SAB). This block's functionality is defined by its capability to adjust the deflection angle, the output of the SGB. The SAB can be considered as a torque amplifier.

As a gearbox, the main function of the SAB is to vary the torque. This block is composed of a set of gears, two ring gears and one spur gear (see Figure 9a). The ring gears are considered as the input carried by a double lever arm system applying a torque. The lever arm system is composed of two drive rods and a prismatic joint L2, as shown in Figures 9a and 2, respectively. The displacement of each driven rod is defined by the parameter a. The actuator M, as depicted in Figure 2b, allows one to change the value of the parameter a acting on the reduction ratio of the SAB. The parameter a defines the distance between the driven rod and the cam center. The position of the driven rod can also be defined by the parameter x when it slides along the ring gear. All these parameters are given in Figure 9b.

V2SOM's deflection angle at the output is θ, Figure 9b, which is the output of the SAB. The relation between the output torque T_θ and the input torque T_γ is given by the following:

$$\frac{T_\theta}{T_\gamma} = \frac{R_1}{R_2}\left(-1 + \frac{\cos\delta}{\sqrt{\left(\frac{a}{R}\right)^2 - \sin\delta^2}}\right)^{-1} : \text{where } \delta = \text{atan}\left(\frac{a\sin\theta}{R - a\cos\theta}\right) \tag{15}$$

$$\begin{cases} R\sin\left(\frac{\pi R_2}{2R_1}\right) \le a \le R \\ |\theta| \in \left[0, \frac{\pi}{2}\left(1 - \frac{R_2}{R_1}\right) - \text{acos}\left(\frac{R}{a}\sin\left(\frac{\pi R_2}{2R_1}\right)\right)\right]. \end{cases} \tag{15a} \tag{15b}$$

Equation (15-a) reports the condition of the tuning parameter a appearing in the rate of the input–output torque equation and defines the torque behavior of the SAB. When the input angle γ changes inside the bounding interval $[-90°, 90]$, the output angle θ is still limited inside the range given by Equation (15-b).

An approximation of Equation (15) can be written when the deflection θ is close to zero value using second-order Taylor polynomial approximation. The obtained equation, noted F_{ideal}, is expressed as follows:

$$F_{ideal}(a, \theta) = \frac{a}{R - a} \tag{16}$$

(a)   **(b)**

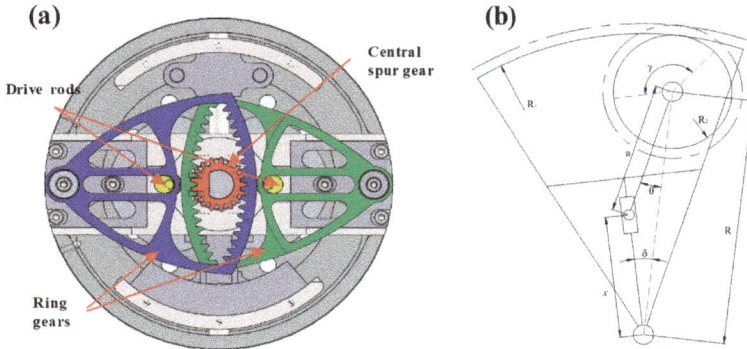

Figure 9. Stiffness adjusting block: (**a**) CAD model, cross-section view' (**b**) simplified scheme with single ring gear.

Figure 10 presents the real as well as ideal curves of the SAB's reduction ratio $\frac{T_\theta}{T_\gamma}$ given by Equations (15) and (16), respectively. The curves are computed for the numerical value $\frac{R_1}{R_2} = 7.5$, considered for illustration purposes. One can observe that in the vicinity of the zero-deflection value, the two curves overlap. This occurs for normal working conditions correlating with range (I) in Figure 8. Beyond the zero-deflection value, the system toggles to the range (II) and the two curves slightly split up. The approximation formula of the reduction ratio, $F_{ideal}(a, \theta)$, meets well the real curves under normal working conditions. In addition, one observes the possible tunability through parameter a. The only drawback lies in the fact that the curves split up, which can be avoided, as explained in the next section, with a correction on the profile of the cam.

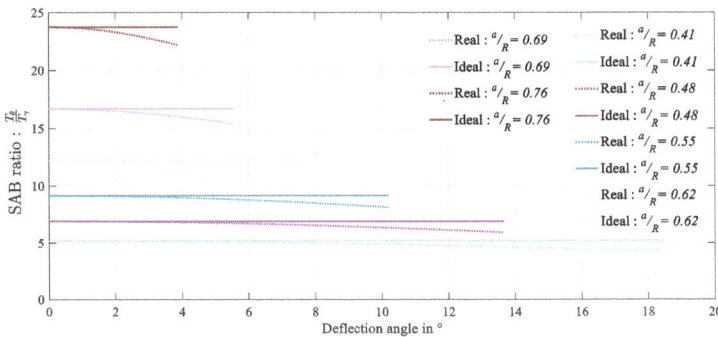

Figure 10. SAB ratio $\frac{T_\theta}{T_\gamma}$ curves (dotted line) and their ideal approximation F_{ideal} (solid line).

3. V2SOM's Prototype

In the previous sections, we presented the two functional blocks and concluded that in order to get the overall ideal approximation of the SAB, the cam's profile must be corrected. This change compensates for the declining trend of the reducer's ratio, thus making the SAB behave as a continuously tunable quasi-linear reducer. The cam's profile originates from the T_γ expression in Equation (14) which is corrected with a second-order polynomial factor, resulting in T_{cor} as follows:

$$T_{cor}(\gamma) = T_\gamma(\gamma) \cdot \left(a_0 + a_1\gamma + a_2\gamma^2 \right) \tag{17}$$

with a_0, a_1 and a_2 are real coefficients. The second-order polynomial factor is chosen in a way that the T_γ value is modified only in the range (II) while the continuity between the two deflection ranges is preserved. This is done with a simple optimization based on the least squares method, where the error is the difference between a set of points representing the real SAB characteristic and their ideal matching set. Figure 11 summarizes the design methodology of V2SOM.

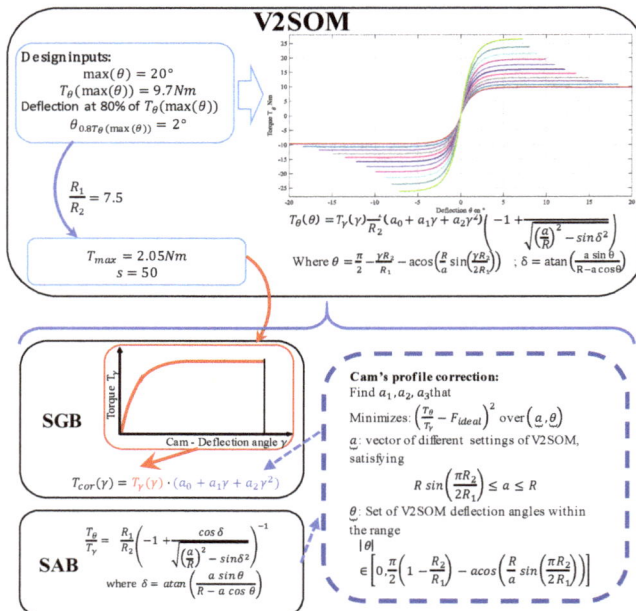

Figure 11. Illustration of V2SOM's design.

The V2SOM prototype was developed with its two functional blocks as illustrated in Figure 12. Two miniature linear motion actuators (Series PQ12, [23]) were used in the upper block. These compact, miniature-sized actuators present the following characteristics: maximal speed (no load), 9 mm/s; stroke, 20 mm; and maximal force, 35 N. The considered parameters of the torque curve vs. deflection, corresponding to the theoretical curve shown in Figure 8, are $T_{max} = 2.05 \, Nm$ and $s = 50$. The real coefficients of the second-order polynomial obtained through optimization process and handled in the cam design are $[a_1, a_2, a_3] = [1.001, -0.0369, 2.588]$.

Figure 13 shows the theoretical curves (solid line) which maintain a relatively constant threshold in the range (II). This indicates the quality of the correction brought to the cam's profile. The experimental curves show a slight deviation from their corresponding theoretical ones that is due to the imperfection of mechanical parts (e.g., natural friction phenomenon). Overall, a practically good match can be concluded as the crucial deflection range (I) shows a good match and the collision range (II) slightly deviates from its theoretical value. The presented V2SOM prototype has a cylindrical volume of 92 mm in diameter and 78 mm in height, as shown in Figure 3.

Figure 12. V2SOM prototype with each two functional blocks: (**a**) SAB: stiffness adjusting block, (**b**) SGB: stiffness generator block.

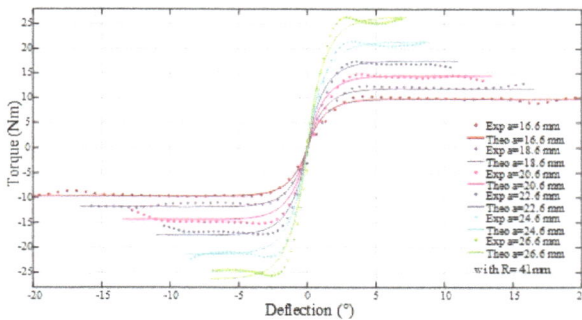

Figure 13. V2SOM's torque vs. deflection theoretical and experimental curves for the eleven different SAB settings.

4. Performance Evaluation: V2SOM vs. Constant Stiffness

In this section, a comparison between V2SOM vs. a tunable constant stiffness profile is carried out on the choice of both HIC and ImpF criteria via simulation, as quantitative evaluation. A mechanical model of HR shock [10,17] is considered and implemented under a Matlab/Simulink platform for this purpose.

4.1. Safety Criteria

The safety of pHRI is quite problematic, particularly in terms of quantification as well as its validity to the whole-body regions. The most widely considered safety criteria include, but are not limited to, the following:

- HIC: this criterion quantifies the high accelerations of brain concussion during blunt shocks even for a short amount of time, for example, HIC_{15} less than 15 ms is sufficient for robotics applications according to [24], which can cause severe irreversible health issues [25].
- ImpF (also known as contact force): this criterion is quite interesting as it can be applied to the whole-body regions. The contact force value is computed for a specific contact surface with a minimum 2.70 cm^2 area.
- Compression criterion (CompC): this criterion reflects a damaging effect of human–robot (HR) collision by means of a deformation depth, mainly considered for the compliant regions such as chest and belly.

The head region is the most critical part of the human body compared to the trunk region which is naturally compliant. The CompC criterion is not relevant for the head region as the skull is quite rigid. In contrast, HIC and ImpF are considered for their complementary aspect of HR shock evaluation. HIC is suitable for internal damage evaluation as it quantifies dangerous brain concussions, while ImpF is suitable for external damage evaluation. The collaborative workspace should be designed, as noted in ISO/TS15066 permits, in a way that free head motion cannot be compromised as the first step to guaranteeing safe pHRI.

4.2. Human–Robot Collision Model

The most critical body region, as investigated in the literature [17,18], is still the human head from the perspective of safety problems. With respect to that investigation, a theoretical modeling of a dummy head hardware in a crash test was proposed and validated experimentally. The mechanical model is shown in Figure 14 and parameterized according to [10]:

- Neck viscoelastic parameters $d_N = 12$ [Ns/m], $k_N = 3300$ [N/m];
- Head's mass $M_{head} = 5.09$ [Kg] and linear displacement x;
- Contact surface viscoelastic parameters $d_c = 10$ [Ns/m], $k_c = 1500$ [N/m];
- Robot arm contact position $l = 0.6$ [m] and inertia $I_{arm} = 0.14$ $\left[\text{Kgm}^2\right]$;
- Rotor inertia I_{rotor}, torque τ_{rotor} and angular position θ_1;
- Stiffness of the variable stiffness mechanism K and angular deflection $\theta = \theta_1 - \theta_2$.

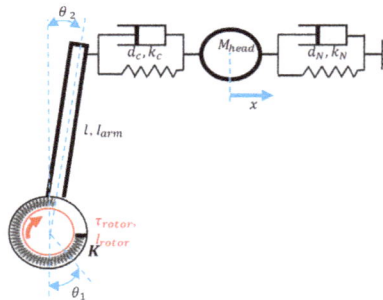

Figure 14. Mechanical model of dummy head hardware collision against a robot arm.

Both criteria, ImpF and HIC, are deduced from the collision model. The first one, ImpF, is obtained from simulation data as the applied force on the contact surface. The second one, HIC, is computed as a result of the following optimization problem:

$$HIC_{15} = \max_{t_1,t_2} \left[\left(\frac{1}{(t_2 - t_1)} \int_{t_1}^{t_2} a(t)dt \right)^{2.5} (t_2 - t_1) \right], \text{ Subject to } t_2 - t_1 \le 15 \text{ ms} \qquad (18)$$

where $a(t)$ is the head acceleration value at instant t.

4.3. Simulation Results of HR Collision

In the ensuing analysis, a comparison between V2SOM and a constant stiffness (CS) VSM was performed through simulation of the HR collision model under Matlab/Simulink. An identical elastic deflection value was considered for CS and V2SOM deflection at 80% of T_{max}. This torque value defines the deflection range of the normal working mode for the V2SOM after which the shock absorbing mode is triggered. The shock absorbing mode is triggered when the torque reaches T_1 value (see

Figure 15). In this case, the springs are compressed by the followers' displacement as a result of the cam rotation (as shown in Figures 5 and 6).

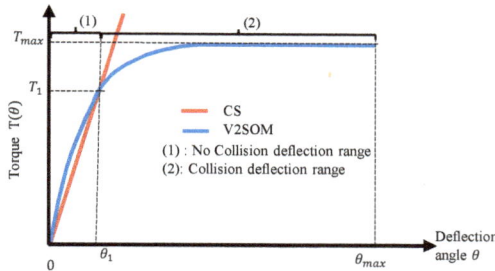

Figure 15. V2SOM and CS profile, (T_{max}, T_1) = (15, 12) [Nm].

The simulation aims to emphasize the decoupling capability of V2SOM along inertia and torque in comparison to an equivalent CS-based variable stiffness mechanism.

Inertia decoupling. The obtained results given in Figure 16 show that V2SOM presents more than an 80% improvement for the HIC criterion compared to CS. On the other hand, an improvement from 10% up to 40% is observed on the ImpF curves. HIC_{V2SOM} and $ImpF_{V2SOM}$ curves are still stable for a large range of rotor inertia. One can conclude from these results that V2SOM presents a high inertia decoupling capability compared to a CS-based variable stiffness mechanism. This characteristic means that in the case of HR collision, the human body sustains only arm side inertia rather than the heavy resulting arm and rotor inertia.

Torque decoupling. The obtained results given in Figure 17 show quasi-constant curves for V2SOM. The variation of motor applied torque τ_{rotor} does not affect the two criteria values, HIC and ImpF. An improvement of 10% up to 40% is observed for V2SOM for the ImpF criterion. This outcome is alleviated by the HIC values which confirm the torque decoupling capacity of a V2SOM similar to elastic behavior.

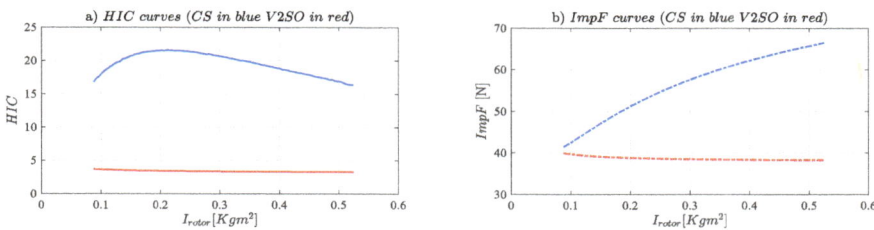

Figure 16. I_{rotor} simulation results; ($\tau_{rotor}, T_{max}, T_1$) = (10, 15, 12) [Nm]; $c = 37$ [SI]; $\dot{\theta}_1 = \pi \left[\text{rads}^{-1}\right]$.

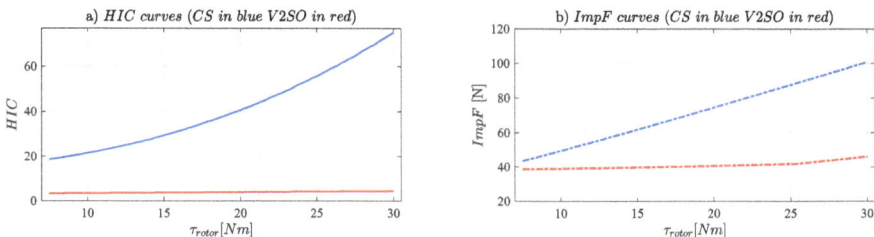

Figure 17. τ_{rotor} simulation results; $I_{rotor} = 0.175 \left[\text{Kgm}^2\right]$; ($\tau_{rotor}, T_{max}, T_1$) = (7.5→30, 15,12) [Nm]; $c = 37$; $\dot{\theta}_1 = \pi \left[\text{rads}^{-1}\right]$.

5. Conclusions

This paper deals with the design methodology of the variable stiffness safety-oriented mechanism (V2SOM). This new device, as its name indicates, comes to ensure the safety of physical human–robot interaction (pHRI) as well as to reduce the dynamics' drawbacks of making robots compliant. Due to its two continuously linked functional modes, high and low stiffness modes, this novel device presents a high inertia decoupling capacity. The V2SOM mechanism's working principle has been presented as well as its two functional blocks. The mechanical description of each block in addition to its mathematical models have been detailed. The interaction along the whole profile response between the two blocks has been discussed. Each block accomplishes a specific role, the first generating the desired stiffness profile through a cam-follower system and the second adjusting the stiffness profile through compact ring gears. The theoretical as well as the preliminary results of the first V2SOM prototype has been presented and discussed. Further experimental results will be addressed in future publications. The performance evaluation of V2SOM in terms of safety through an evaluation of safety criteria was performed. The impact force (ImpF) criterion and the head injury criterion (HIC) for external and internal damage evaluation of blunt shocks were considered, respectively.

Currently, a faster and lighter version of this device is under development, knowing that the current version weighs about 970 g with all its integrated control electronics.

Author Contributions: Conceptualization, Y.A. and M.A.L.; Data curation, Y.A. and M.A.L.; Formal analysis, Y.A.; Funding acquisition, M.A. and S.Z.; Investigation, Y.A., M.A.L., M.A. and S.Z.; Methodology, Y.A. and M.A.L.; Project administration, M.A.; Software, Y.A. and M.A.L.; Supervision, M.A.L. and S.Z.; Validation, Y.A., M.A.L., M.A. and S.Z.; Writing—original draft, Y.A. and M.A.L.; Writing—review & editing, Y.A. and M.A.L.

Funding: This research was funded by the French National Research Agency, convention ANR-14-CE27-0016, under the ANR project SISCob "Safety Intelligent Sensor for Cobots". This research was also supported by the French region "Nouvelle-Aquitaine" (program HABISAN 2015-2020) with the financial participation of the European Union (FEDER/ERDF, European Regional Development Fund).

Acknowledgments: This work was sponsored by the French government research program *Investissements d'avenir* through the Robotex Equipment of Excellence (ANR-10-EQPX-44).

Conflicts of Interest: The authors declare no conflict of interest.

References

1. Khatib, O.; Yokoi, K.; Brock, O.; Chang, K.; Casal, A. Robots in human environments: Basic autonomous capabilities. *Int. J. Robot. Res.* **1999**, *18*, 684–696. [CrossRef]
2. Tobe, F. Why Co-Bots Will Be a Huge Innovation and Growth Driver for Robotics Indus-try. Available online: http://spectrum.ieee.org/automaton/robotics/industrial-robots/collaborative-robots-innovation-growth-driver (accessed on 30 December 2015).
3. De Santis, A.; Siciliano, B.; De Luca, A.; Bicchi, A. An atlas of physical human–robot interaction. *Mech. Mach. Theory* **2008**, *43*, 253–270. [CrossRef]
4. Park, J.J.; Haddadin, S.; Song, J.B.; Albu-Schäffer, A. Designing optimally safe robot surface properties for minimizing the stress characteristics of human-robot collisions. In Proceedings of the 2011 IEEE International Conference on Robotics and Automation, Shanghai, China, 9–13 May 2011; pp. 5413–5420.
5. Fritzsche, M.; Elkmann, N.; Schulenburg, E. Tactile sensing: A key technology for safe physical human robot interaction. In Proceedings of the 6th International Conference on Human-Robot Interaction (HRI'11), Lausanne, Switzerland, 6–9 March 2011; ACM: New York, NY, USA, 2011; pp. 139–140.
6. Chiaverini, S.; Siciliano, B.; Villani, L. A survey of robot interaction control schemes with experimental comparison. *IEEE/ASME Trans. Mech.* **1999**, *4*, 273–285. [CrossRef]
7. Bicchi, A.; Tonietti, G.; Bavaro, M.; Piccigallo, M. Variable stiffness actuators for fast and safe motion control. *Robot. Res.* **2005**, 527–536.
8. Grioli, G.; Wolf, S.; Garabini, M.; Catalano, M.; Burdet, E.; Caldwell, D.; Carloni, R.; Friedl, W.; Grebenstein, M.; Laffranchi, M.; et al. Variable stiffness actuators: The user's point of view. *Int. J. Rob. Res.* **2015**, *34*, 727–743. [CrossRef]

9. Jianbin, H.; Zongwu, X.; Minghe, J.; Zainan, J.; Hong, L. Adaptive Impedance-controlled Manipulator Based on Collision Detection. *Chin. J. Aeronaut.* **2009**, *22*, 105–112. [CrossRef]
10. Gao, D.; Wampler, C.W. Assessing the Danger of Robot Impact. *IEEE Robot. Autom. Mag.* **2009**, *16*, 71–74. [CrossRef]
11. Pratt, G.A.; Williamson, M.M. Series elastic actuators. In Proceedings of the 1995 IEEE/RSJ International Conference on Intelligent Robots and Systems, Human Robot Interaction and Cooperative Robots, Pittsburgh, PA, USA, 5–9 August 1995; Volume 1, pp. 399–406.
12. Mathijssen, G.; Cherelle, P.; Lefeber, D.; Vanderborght, B. Concept of a Series-Parallel Elastic Actuator for a Powered Transtibial Prosthesis. *Actuators* **2013**, *2*, 59–73. [CrossRef]
13. Zinn, M.; Roth, B.; Khatib, O.; Salisbury, J.K. A New Actuation Approach for Human Friendly Robot Design. *Int. J. Rob. Res.* **2004**, *23*, 379–398. [CrossRef]
14. Wolf, S.; Grioli, G.; Eiberger, O.; Friedl, W.; Grebenstein, M.; Höppner, H.; Burdet, E.; Caldwell, D.G.; Carloni, R.; Catalano, M.G.; et al. Variable Stiffness Actuators: Review on Design and Components. *IEEE/ASME Trans. Mechatron.* **2016**, *21*, 2418–2430. [CrossRef]
15. Ayoubi, Y.; Laribi, M.A.; Courrèges, F.; Zeghloul, S.; Arsicault, M. A Complete Methodology to Design a Safety Mechanism for Prismatic Joint Implementation. *IEEE/RSJ Int. Conf. Intell. Robot. Syst.* **2016**, 304–309.
16. Ayoubi, Y.; Laribi, M.A.; Zeghloul, S.; Arsicault, M. Design of V2SOM: The Safety Mechanism for Cobot's Rotary Joints. In *Mechanism Design for Robotics. MEDER 2018. Mechanisms and Machine Science*; Gasparetto, A., Ceccarelli, M., Eds.; Springer: Cham, The Netherlands, 2019; Volume 66.
17. López-Martínez, J.; García-Vallejo, D.; Giménez-Fernández, A.; Torres-Moreno, J.L. A Flexible Multibody Model of a Safety Robot Arm for Experimental Validation and Analysis of Design Parameters. *J. Comput. Nonlinear Dyn.* **2013**, *9*, 1–9. [CrossRef]
18. Flacco, F.; de Luca, A. Residual-based stiffness estimation in robots with flexible transmissions. In Proceedings of the 2011 IEEE International Conference on Robotics and Automation, Shanghai, China, 9–13 May 2011; pp. 5541–5547.
19. Cirillo, A.; de Maria, G.; Natale, C.; Pirozzi, S. A mechatronic approach for robust stiffness estimation of variable stiffness actuators. In Proceedings of the 2013 IEEE/ASME International Conference on Advanced Intelligent Mechatronics, Wollongong, Australia, 9–12 July 2013; pp. 399–404.
20. Ayoubi, Y.; Laribi, M.A.; Arsicault, M.; Zeghloul, S.; Courreges, F. *Mechanical Device with Variable Compliance for Rotary Motion Transmission*, FR/IFBT17CNRCOB. 2017; France.
21. Petit, F.; Friedl, W.; Hannes, H.; Grebenstein, M. Antagonistic Variable Stiffness Mechanism. *Trans. Mechatron.* **2015**, *20*, 684–695. [CrossRef]
22. Eiberger, O.; Haddadin, S.; Weis, M.; Albu-Schäffer, A.; Hirzinger, G. On joint design with intrinsic variable compliance: Derivation of the DLR QA-joint. In Proceedings of the 2010 IEEE International Conference on Robotics and Automation, Anchorage, Alaska, 3–8 May 2010; pp. 1687–1694.
23. Hyun, D.; Yang, H.S.; Park, J.; Shim, Y. Variable stiffness mechanism for human-friendly robots. *Mech. Mach. Theory* **2010**, *45*, 880–897. [CrossRef]
24. Bicchi, A.; Tonietti, G. Fast and 'soft-arm' tactics. *IEEE Robot. Autom. Mag.* **2004**, *11*, 22–33. [CrossRef]
25. Firgelli. Available online: http://www.firgelli.com (accessed on 5 January 2019).

![robotics logo] *robotics*

MDPI

Article

A Toolbox for the Analysis of the Grasp Stability of Underactuated Fingers [†]

Giovanni Antonio Zappatore [1], Giulio Reina [2,*] and Arcangelo Messina [2]

[1]   BionIT Labs Company, Via A. Bortone, 73100 Lecce, Italy; g.zappatore@gmail.com
[2]   Dipartimento di Ingegneria dell'Innovazione, Università del Salento, Via Arnesano, 73100 Lecce, Italy; arcangelo.messina@unisalento.it
*   Correspondence: giulio.reina@unisalento.it; Tel.: +39-0832297814
†   This paper is an extended version of our paper published in Zappatore, G.; Reina, G.; Messina, A. A proposed software framework for studying the grasp stability of underactuated fingers. In Proceedings of the 4th IFToMM Symposium on Mechanism Design for Robotics, Udine, Italy, 11–13 September 2018; pp. 202–210.

Received: 29 December 2018; Accepted: 28 March 2019; Published: 6 April 2019

Abstract: In the design of humanoid robotic hands, it is important to evaluate the grasp stability, especially when the concept of underactuation is involved. The use of a number of degrees of actuation lower than the degrees of freedom has shown some advantages compared to conventional solutions in terms of adaptivity, compactness, ease of control, and cost-effectiveness. However, limited attention has been devoted to the analysis of grasp performance. Some specific issues that need to be further investigated are, for example, the impact of the geometry of the fingers and the objects to be grasped and the value of the driving mechanical torques applied to the phalanges. This research proposes a software toolbox that is aimed to support a user towards an optimal design of underactuated fingers that satisfies stable and efficient grasp constraints.

Keywords: humanoid robotic hands; underactuated fingers; graphical user interface; grasp stability

1. Introduction

In the last few years, increasing interest has been devoted towards compliant and underactuated hands as a compact, reliable, and flexible grasping solution in manipulation applications [1,2]. However, relatively limited attention has been given to the development of simulation tools that address the specific challenges connected with underactuated grasping [3]. Notable examples are GraspIt! [4] and OpenGrasp [5]. Both simulators allow a set of common objects and various types of grippers to be analyzed, but they are not well suited for grasp stability analysis, especially when underactuated architectures are considered. Other recent efforts that address the specific design of under-actuated hands include [6], whereas SynGrasp [7] is a MATLAB toolbox for grasp analysis of fully or underactuated robotic hands. Finally, in [8], the Yale-CMU-Berkeley (YCB) object and model set is presented that is intended to be used to facilitate benchmarking in robotic manipulation, prosthetic design, and rehabilitation research.

This paper introduces a simulation toolbox that the authors developed during their current efforts at BionIT Labs towards an efficient design of Adam's Hand: a transradial myoelectric prosthesis that uses a highly underactuated mechanism, composed of 14 differential stages actuated by a single motor −15 degrees of freedom (DOFs), 1 degree of actuation (DOA) [9]. The underactuation among the fingers is obtained by symmetrically stacking five bevel gear differential stages, while the underactuation within each finger is obtained by stacking serially two differential idler pulleys per finger. A functional scheme of the designed mechanism is shown in Figure 1, while a more in-depth analysis of the proposed mechanism can be found in [10,11]. This paper extends the study, preliminary presented by the authors in [12], on the contact forces generated by the underactuated fingers during enveloping grasps. The overall goal is to optimize their features by maximizing the contact conditions for which a stable grasp can be achieved. As explained in the following section, these contact situations are identified by a combination of phalanx flexion/extension angles and contact points of the phalanges with the object to be grasped. In order to simplify the analysis that involves a high number of variables, a software is presented in Section 4 to support a user during the design stage. The software framework, available upon request, is developed in the Mathematica environment, which is a very powerful symbolic language and well-established in the scientific and industrial world. Mathematica programming environment allows to easily exploit other specific tools and built-in math functions enabling the exploration of multiple approaches and the integration with other analysis tools, e.g., statistical processing of experimental data, optimization, dynamic models, and simulations.

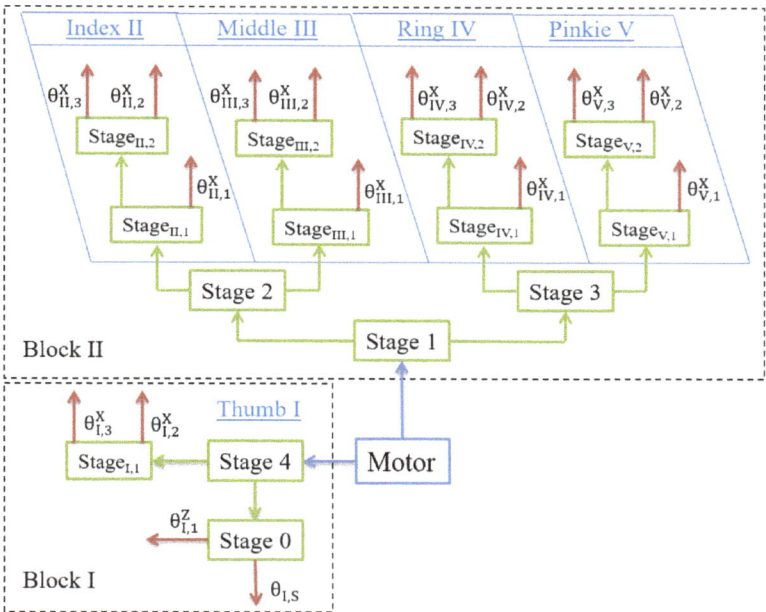

Figure 1. The Adam's Hand mechanism employs the following underactuation tree [11].

2. General Static Model

Drawing on [13], the model of underactuated finger used in this paper is shown in Figure 2. The following assumptions hold: the finger motion is planar (no abduction/adduction), and all the n phalanges, which are driven by a single actuator, are linked through revolute joints. Equating the input and output virtual powers of this system, one obtains:

$$t^T \omega_a = \sum_{i=1}^{n} \xi_i \circ \zeta_i \tag{1}$$

where:

- t is the input torque vector exerted by the actuator (T_a) and the springs located between the phalanges ($T_2, \ldots, T_n$):

$$t = \begin{pmatrix} T_a & T_2 & \ldots & T_n \end{pmatrix}^T \tag{2}$$

- ω_a is the corresponding joint velocity vector:

$$\omega_a = \begin{pmatrix} \dot{\theta}_a & \dot{\theta}_2 & \cdots & \dot{\theta}_n \end{pmatrix}^T \tag{3}$$

where θ_i is the ith joint variable.

- ξ_i is the twist of the ith contact point on the ith phalanx (assuming one contact per phalanx) with a corresponding wrench ζ_i, and the operator "$\circ$" stands for the reciprocal product of screws in the plane.

It can be shown that:

$$\xi_i \circ \zeta_i = f^T(J\dot{\theta}) = f^T(JT\omega_a) \tag{4}$$

where:

- f^T is the vector of the resultant of contact forces, f_i, normal to phalanx $1, \ldots, n$:

$$f = \begin{pmatrix} f_1 & f_2 & \cdots & f_n \end{pmatrix}^T \tag{5}$$

- J is the *Jacobian Matrix*, a $n \times n$ square matrix which depends only on the location of the contacts k_i on the phalanges and their relative orientation r_{ij}^T, their length l_i and the friction coefficients μ_i and η_i:

$$J = \begin{bmatrix} k_1 + \eta_1 & 0 & \cdots & 0 \\ r_{12}^T(x_2 - \mu_2 y_2) + \eta_2 & k_2 + \eta_2 & \cdots & 0 \\ \cdots & \cdots & \cdots & \cdots \\ r_{1n}^T(x_n - \mu_n y_n) + \eta_n & r_{2n}^T(x_n - \mu_n y_n) + \eta_n & \cdots & k_n + \eta_n \end{bmatrix} \tag{6}$$

- T is the *Transmission Matrix*, a $n \times n$ square matrix which depends on the stage transmission ratios x_i of the mechanism used to propagate the actuation torque to the phalanges:

$$T = \begin{bmatrix} 1 & -x_1 x_2 & -x_1 x_2 x_3 & \cdots & -\prod_{i=1}^{n} x_i \\ 0_{n-1}^T & & I_{n-1} & \end{bmatrix} \tag{7}$$

where I_{n-1} and 0_{n-1}^T are the identity matrix and the zero vector of dimension $(n-1)$.

Then, considering Equations (1) and (4), the equilibrium of virtual power for the system results:

$$t^T \omega_a = f^T(JT\omega_a) \tag{8}$$

from which one obtains a useful relationship between the actuator torques and the contact forces:

$$f = J^{-T} T^{-T} t \tag{9}$$

It should be noted that a n-output m-input underactuated mechanism requires $n - m$ springs in order to be statically determined. For this reason, depending on the mechanism design, it is possible that a torsion spring will be required also in the base joint O_1. In this case, matrix J remains the same, while vector t and matrix T^{-1} respectively become:

$$t = \begin{bmatrix} T_a & T_1 & T_2 & \cdots & T_3 \end{bmatrix}^T \tag{10}$$

$$T^{-1} = T^* = \begin{bmatrix} 1 & x_1 x_2 & x_1 x_2 x_3 & \cdots & \prod_{i=1}^{n} x_i \\ & & I_n & \end{bmatrix} \tag{11}$$

where I_n is the identity matrix of dimension n, and T^* is now a rectangular matrix of dimensions $(n + 1) \times n$. Equation (9) in this case becomes:

$$f = J^{-T}T^{*T}t \tag{12}$$

and the forces obtained are the same calculated in absence of the base joint spring, except for f_1, which contains the additional term $T_1 \dfrac{1}{k_1 + \eta_1}$ (and this holds for any number of phalanges). This result represents the most general one, since if $T_1 = 0$ also f_1 equals the one previously obtained. For this reason, from now on, matrices J (Equation (6)) and T^* (Equation (11)) and vector t (Equation (10)) will be used to optimize the fingers design using Equation (12).

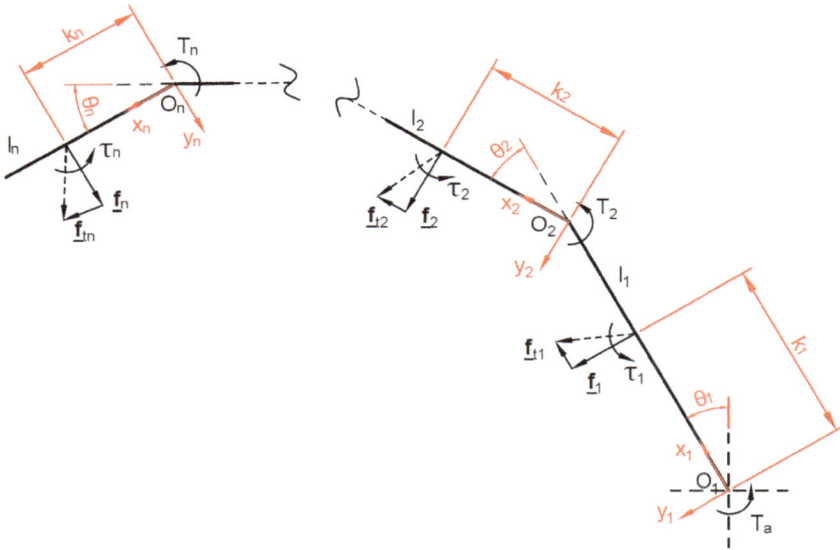

Figure 2. Model of an underactuated finger.

2.1. Impact of Phalanx Thickness

As can be seen in Figure 3, when the *i*th phalanx thickness ϵ_i is not negligible the angle θ_i should be augmented by the quantity:

$$\psi_i = \arctan \frac{\epsilon_i}{k_i} \tag{13}$$

and the contact location k_i should be shifted by

$$k_i^* = \sqrt{k_i^2 + \epsilon_i^2} \tag{14}$$

In addition, when friction is non-zero, the equilibrium locus changes due to the moment generated by the tangential force, which can be modelled using the coefficient η_i: the tangential force produces a moment about O_i equal to $-f_{ti}\epsilon_i$. This moment can be seen as a wrench with the same normal and tangent forces and a torque τ_i equal to the case of a zero thickness phalanx. Therefore, one gets $\tau_i = \eta_i f_i = -f_{ti}\epsilon_i = -f_i\mu_i\epsilon_i$, thus the equivalent instantaneous rolling friction coefficient is $\eta_i = -\epsilon_i\mu_i$. The latter coefficient must be added to the previous value of η_i describing the contact friction (even if it is zero). This change can be reflected directly into the matrix J to obtain the new force expressions.

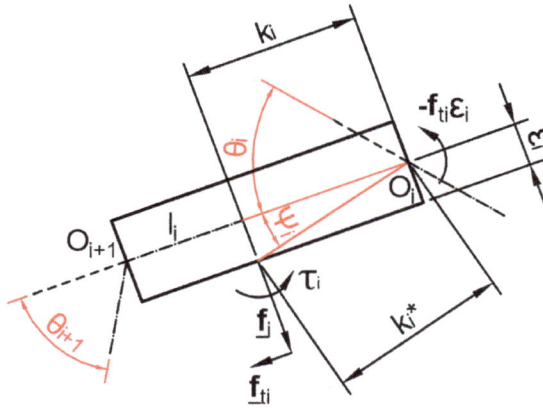

Figure 3. Impact of the phalanx thickness.

2.2. Positive Definiteness of the Forces

Given a set of geometric parameters, Equation (9) or (12) provides the contact configurations defined by the pair (k^*, θ^*)

$$k^* = \begin{bmatrix} k_2 \\ k_3 \\ \dots \\ k_n \end{bmatrix} \quad and \quad \theta^* = \begin{bmatrix} \theta_2 \\ \theta_3 \\ \dots \\ \theta_n \end{bmatrix} \tag{15}$$

that ensure full positiveness of the vector f. The set of these contact situations corresponds to the stable part of the space spanned by the contact situations pair (k^*, θ^*) which are referred to as the *space of contact configurations* or *grasp-state space*. Stable grasps correspond to contact situation pairs for which the vector f has no negative component, that is, the phalanges in contact with an object have a positive (or zero) contact forces. The other phalanges that are not in contact with the object must correspond to zero contact forces. It should be underlined that this approach tries to characterize the finger itself, independently from the object being grasped.

It should be also considered that the grasps requiring all the phalanges correspond only to a subset of all the possible grasps: fewer-than-n-phalanges grasps can also be stable if each phalanx which contacts the object has a strictly positive contact force and each phalanx not in contact with the object has a null contact force.

3. Contact Forces Writing for the Proposed Finger Mechanism

The general equations presented in Section 2 are written for the scheme proposed in Figure 1, for both the two-phalanx thumb (I) and for the four three-phalanx fingers, from index (II) to pinkie (V). As mentioned in Section 1, the prosthetic hand under study features $n = 15$ DOFs that are actuated by just $m = 1$ DOA, so $n - m = 15 - 1 = 14$ springs are required to solve the static equilibrium equations. Due to symmetry considerations, these springs have to be located in all the joints of each finger (three springs for finger $II \div V$ and two springs for finger I), so that also the base joint (O_1) of each finger will be linked through a spring to the fixed palm.

3.1. Two-Phalanx Finger

Three DOFs are assigned to the thumb, corresponding to proximal and distal phalanges flexion/extension and to metacarpus abduction/adduction. These members are interconnected via three revolute joints. A torsional spring that is positioned in each joint, which links the phalanges, the metacarpus, and the palm. Since the analysis carried out in Section 1 does not consider out-of-the-flexion-plane movements, in the following analysis the metacarpus motion is constrained so that the thumb results as composed only by the proximal and the distal phalanges. Whereas this approximation subtracts generality to the analysis, it should be considered that many grasp typologies can be obtained with the metacarpus fixed relatively to the palm. The model of the thumb is presented in Figure 4a.

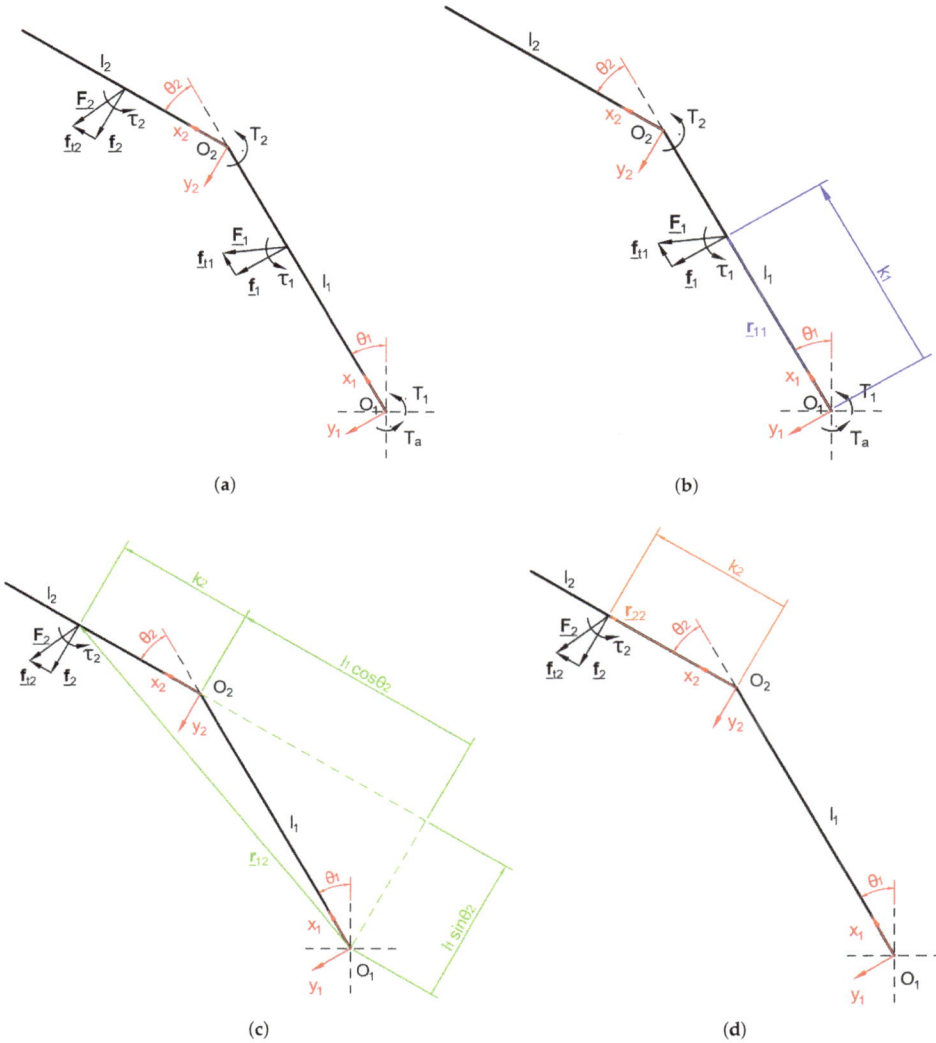

Figure 4. Model of the underactuated 2-phalanx thumb (**a**) and determination of J_l elements: vectors r_{11} (**b**), r_{12} (**c**), and r_{22} (**d**).

Both flexion/extension of the two phalanges and abduction/adduction of the metacarpus are driven by the torque deriving from the bevel gear differential stage 4, as shown in Figure 1.

Matrix J_I according to Equation (6) is given by:

$$J_I = \begin{bmatrix} k_1 + \eta_1 & 0 \\ k_2 + l_1(\cos\theta_2 - \mu_2\sin\theta_2) & k_2 + \eta_2 \end{bmatrix} \tag{16}$$

The physical meaning of J_I is showed in Figure 4b–d.

When phalanx thickness ϵ_i is taken into account, the finger model becomes that shown in Figure 5a, as discussed in Section 2.1.

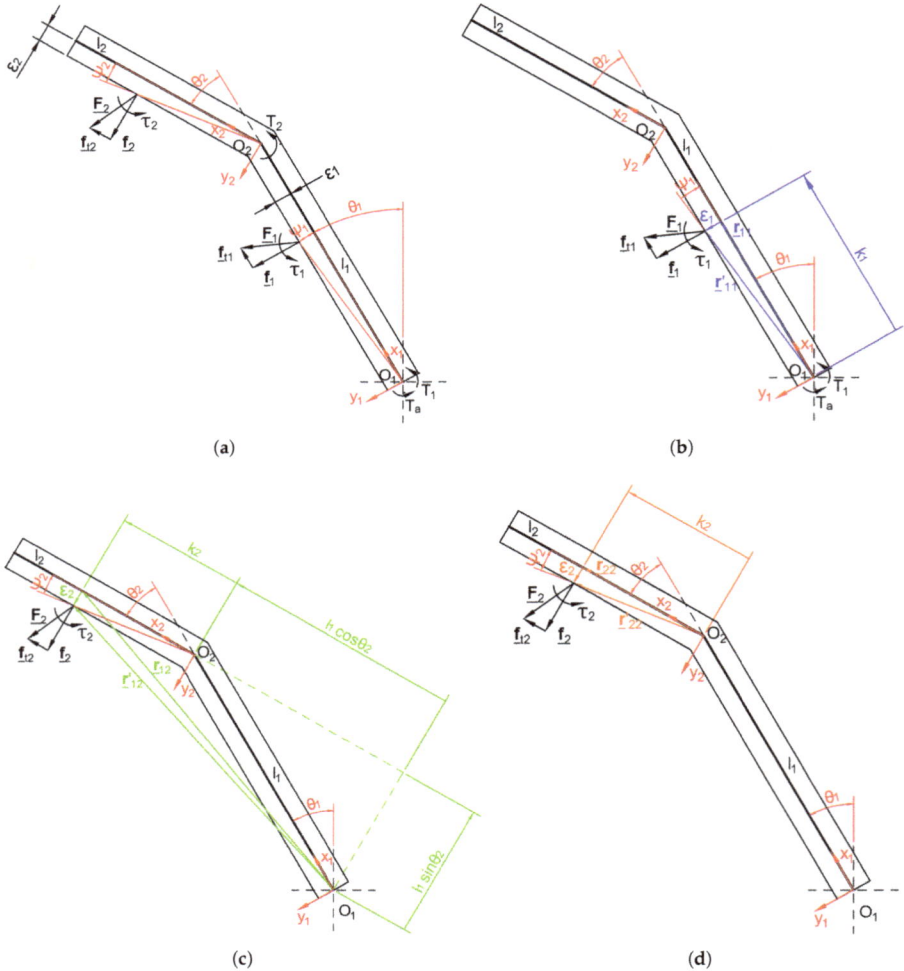

(a)

(b)

(c)

(d)

Figure 5. Model of the underactuated 2-phalanx thumb considering non-negligible phalanx thickness ϵ_1, ϵ_2 (**a**) and determination of the components of vectors r'_{11} (**b**), r'_{12} (**c**), and r'_{22} (**d**).

Matrix J_I gets:

$$J_I = \begin{bmatrix} k_1 + \eta_1 & 0 \\ k_2 + l_1(\cos\theta_2 - \mu_2\sin\theta_2) + \mu_2\eta_2 & k_2 + \eta_2 \end{bmatrix} \tag{17}$$

since vectors r_{ki} in this case acquire another component proportional to phalanx thickness ϵ_i along y_i axis, as shown in Figure 5b–d. This matrix represents a generalization of the matrix reported in Equation (16) for $\epsilon_i \neq 0$ ($i = 1, 2$). Therefore in the following analysis, matrix J_I will be derived from Equation (17).

Matrix T_I and vector t_I can be obtained respectively from Equations (7) and (10):

$$T_I^* = \begin{bmatrix} 1 & x_{12,I} \\ 1 & 0 \\ 0 & 1 \end{bmatrix} \tag{18}$$

$$t_I = \begin{bmatrix} T_{a,I} & T_{1,I} & T_{2,I} \end{bmatrix}^T \tag{19}$$

with $x_{12,I}$ being the transmission ratio between the base and the middle idler pulleys and $T_{a,I}$ the torque exerted by one of the two sun gears of the bevel gear differential Stage 4. Moreover:

$$T_{h,I} = \begin{cases} K_{h,I}(\pi/2 - \theta_{h,I} + Z_{h,I}) & \text{if spring opposes hand opening} \\ -K_{h,I}(\theta_{h,I} + Z_{h,I}) & \text{if spring opposes hand closing} \end{cases} \tag{20}$$

with $K_{h,I}$ being the spring stiffness and $Z_{h,I}$ the spring preload for joints $h = 1, 2$ of the thumb.

The contact forces obtained from Equation (12) are then:

$$\begin{cases} f_1 = (T_1 + T_a)\dfrac{1}{k_1 + \eta_1} - (T_2 + T_a x_{12,I})\dfrac{[k_2 + \eta_2 + \mu_2\epsilon_2 + l_1(\cos\theta_2 - \mu_2\sin\theta_2)]}{(k_1 + \eta_1)(k_2 + \eta_2)} \\ f_2 = (T_2 + T_a x_{12,I})\dfrac{1}{k_2 + \eta_2} \end{cases} \tag{21}$$

Neglecting friction ($\mu_i = \eta_i = 0$ $\forall i$) and considering $x_{12,I} = 1$, these equations become much simpler:

$$\begin{cases} f_1 = (T_1 + T_a)\dfrac{1}{k_1} - (T_2 + T_a)\dfrac{k_2 + l_1\cos\theta_2}{k_1 k_2} \\ f_2 = (T_2 + T_a)\dfrac{1}{k_2} \end{cases} \tag{22}$$

The grasp is stable only if $f_i > 0$ for $i = 1, 2$. By studying the contact situations defined by the pair (k_2, θ_2) for a determined set of geometric, static, and dynamic parameters (phalanx length and thickness, friction coefficients, springs stiffness and preload, actuation torque, ...) the portion of the *grasp-state space* in which all the forces are positive can be obtained. By varying the design parameters, this portion can be maximized in order to ensure a stable grasp for the largest number of contact situation achievable.

3.2. Three-Phalanx Fingers

In this case, three DOFs are assigned to each one of fingers $II \div V$. They all feature three phalanges linked through three revolute joints among them and to the fixed palm. A torsional spring is located in each joint. The model of this finger is presented in Figure 6a. The model is modified as shown in Figure 6b when phalanx thickness is not negligible.

Each finger is driven by the torque delivered by the bevel gear differential stages 2 (fingers II and III) or 3 (fingers IV and V), as shown in Figure 1. Considering the same assumptions made for the two-phalanx finger, the contact forces obtained from Equation (12) are:

$$
\begin{cases}
f_1 = (T_1 + T_a)\dfrac{1}{k_1 + \eta_1} - (T_2 + T_a x_{12})\dfrac{[k_2 + \eta_2 + \mu_2\epsilon_2 + l_1(\cos\theta_2 - \mu_2\sin\theta_2)]}{(k_1 + \eta_1)(k_2 + \eta_2)} \\[2mm]
\quad + (T_3 + T_a x_{12} x_{23})\dfrac{[\mu_2\epsilon_2 + l_1(\cos\theta_2 - \mu_2\sin\theta_2)][k_3 + \eta_3 + \mu_3\epsilon_3 + l_2(\cos\theta_3 - \mu_3\sin\theta_3)]}{(k_1 + \eta_1)(k_2 + \eta_2)(k_3 + \eta_3)} \\[2mm]
\quad - (T_3 + T_a x_{12} x_{23})\dfrac{l_1[\cos(\theta_2 + \theta_3) - \mu_3\sin(\theta_2 + \theta_3)]}{(k_1 + \eta_1)(k_3 + \eta_3)} \\[2mm]
f_2 = (T_2 + T_a x_{12})\dfrac{1}{k_2 + \eta_2} - (T_3 + T_a x_{12} x_{23})\dfrac{[k_3 + \eta_3 + \mu_3\epsilon_3 + l_2(\cos\theta_3 - \mu_3\sin\theta_3)]}{(k_2 + \eta_2)(k_3 + \eta_3)} \\[2mm]
f_3 = (T_3 + T_a x_{12} x_{23})\dfrac{1}{k_3 + \eta_3}
\end{cases}
\tag{23}
$$

Again, neglecting friction ($\mu_i = \eta_i = 0\ \forall i$) and considering $x_{12} = x_{23} = 1$, these equations become much simpler:

$$
\begin{cases}
f_1 = (T_1 + T_a)\dfrac{1}{k_1} - (T_2 + T_a)\dfrac{k_2 + l_1\cos\theta_2}{k_1 k_2} \\[2mm]
\quad + (T_3 + T_a)\dfrac{[k_3 l_1\cos\theta_2 + l_1 l_2\cos\theta_2\cos\theta_3 - k_2 l_1\cos(\theta_2 + \theta_3)]}{k_1 k_2 k_3} \\[2mm]
f_2 = (T_2 + T_a)\dfrac{1}{k_2} - (T_3 + T_a)\dfrac{k_3 + l_2\cos\theta_3}{k_2 k_3} \\[2mm]
f_3 = (T_3 + T_a)\dfrac{1}{k_3}
\end{cases}
\tag{24}
$$

The grasp is stable only if $f_i > 0$ for $i = 1, 2, 3$. As for two-phalanx finger, by studying the contact situations defined by the pair $\left(\begin{bmatrix} k_2 \\ k_3 \end{bmatrix}, \begin{bmatrix} \theta_2 \\ \theta_3 \end{bmatrix} \right)$ one can obtain the portion of the *grasp-state space* in which all the forces are positive.

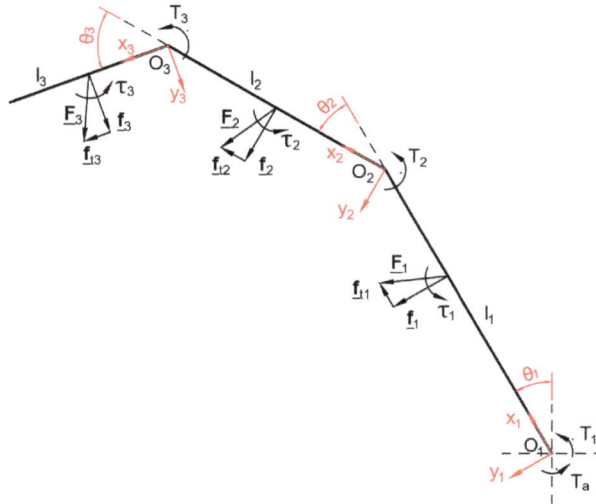

(a)

Figure 6. *Cont.*

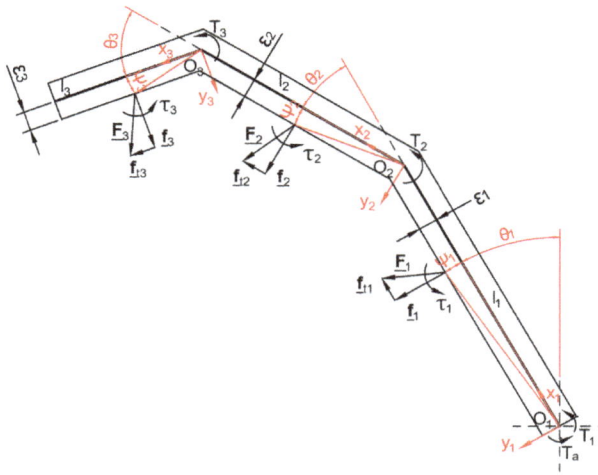

(b)

Figure 6. Model of the underactuated 3-phalanx finger studied, respectively considering negligible (**a**) and non-negligible (**b**) phalanx thickness.

4. Proposed Software

The software toolbox provides a useful tool to simplify the grasp-state space analysis and parametric optimization. It was developed under the Wolfram Mathematica environment [14]. The graphical user interface (GUI) is shown in Figure 7. It consists of four main areas. The first row (2-phalanx finger) and the second row (3-phalanx finger) refer to the finger geometric parameters explained in the previous sections; the third row collects the parameters relative to the target object (position, size, shape), other parameters used to define the limits of the grasp-state space, and the constraints on the normal and tangential contact forces both for two- and three-phalanx fingers; the fourth row includes the graphs that represents the grasp-state space (k_2, θ_2) and the scheme of the two-phalanx finger on the left side, whereas two graphs representing the grasp-state spaces (θ_2, θ_3) and (k_2, k_3) and the scheme of the three-phalanx finger are shown on the right side. Each subsection of the GUI is explained in more detail in the remainder of the paper.

- *Sections 1 and 2—Phalanx length and semi-thickness*

  Phalanx lengths can be set by matching those of the human hand, using standard biomechanical measurements ([15]), as shown in Table 1. Phalanx thickness, instead, has been set considering the size of the mechanical transmission, in order to simplify the following design validation and due to the lack of standard biomechanical measurements in the Literature.
  Note that the software requires as input the phalanx semi-thickness.

Table 1. Standard phalanx lengths [mm].

Phalanx	Thumb (*I*)	Index (*II*)	Middle (*III*)	Ring (*IV*)	Pinkie (*V*)
dp	21.67 ± 1.60	15.82 ± 2.26	17.40 ± 1.85	17.30 ± 2.22	15.96 ± 2.45
mp	-	22.38 ± 2.51	26.33 ± 3.00	25.65 ± 3.29	18.11 ± 2.54
pp	31.57 ± 3.13	39.78 ± 4.94	44.63 ± 3.81	41.37 ± 3.87	32.74 ± 2.77

- *Sections 3, 4, and 5—Springs presence, direction, stiffness, and preload*

  The software foresees the adoption of a spring in each joint; when a joint (e.g., O_1, O_2, O_3) is checked, the rotary spring is activated and the relative stiffness and preload can be set, otherwise the spring is neglected. The user can also choose if the spring opposes opening or closing of the prosthetic hand.

  As stated before, the proposed mechanism requires a spring in each joint of the finger in order to obtain a statically determined finger.

- *Section 6—Friction coefficients*

  The user can choose whether or not to consider friction by checking or unchecking the relative button; in the first case the value of friction coefficients μ_i and η_i ($i = 1, 2, 3$) can be set. These values depend on the material of the object–finger contact: typical values are 0.8 for a steel–steel contact and $1 \div 4$ for solid–rubber (in both cases clean and non-lubricated [16]). As a matter of fact, it should be considered that robotic finger surfaces can be coated with a rubber-like layer to increase friction or indirectly through the use of a tactile sensing device.

  It should be also noted that for a given value of $\mu_{i,static}$ two values of μ_i should be considered, each corresponding to one sliding direction, i.e., $\mu_i = +\mu_{i,static}$ or $\mu_i = -\mu_{i,static}$.

- *Section 7—Torque and transmission ratios*

  The base joint actuation torque T_a must be provided in order to calculate contact forces. Specifically, in the case of fingers $II - V$, this torque is found under the assumption that in the steady-state condition it is equally distributed among all fingers.

  The user can also choose the value of transmission ratios between the phalanges: in order to simplify and speed up the mechanism prototyping, the current version presents unitary transmission ratios.

- *Sections 8 and 9—Force application points and flexion angles*

  The parameters adopted to study the grasp-state space are the phalanx flexion/extension angles θ_2, θ_3 and the force application points, expressed as a percentage of the phalanx length ($k_2 \equiv \%l_2$ and $k_3 \equiv \%l_3$):

  - for two-phalanx fingers this space is of dimensions 3, therefore easily readable on a single $3D$ graph parameterized as a function of (θ_2, k_2);
  - for three-phalanx fingers, instead, at least two different graphs should be considered: the current version of the software shows the force vector components as a function of (θ_2, θ_3) in the first graph (on the left of Figure 7) and as a function of (k_2, k_3) in the second graph (on the right) of the same figure. However, other parameters combinations, such as (θ_2, k_2) or (θ_3, k_3) are easily implementable.

- *Section 10—Grasped object parameters*

  When the object button is checked, the software working modality is affected: force application points, in this case, are automatically defined by the intersection between the phalanges and the object outer shape. The user can choose the object dimension, shape, and position relative to the finger base joint O_1.

- *Section 11—Graphic settings*

  The sliders in this section help defining the grasp-state space boundaries both in terms of (θ_2, θ_3) and (k_2, k_3). They also define the number of points for which numeric integration of a performance index is performed. This index indicates the percentage of the defined grasp-state space, which allows for a stable grasp. The boundaries for contact forces can also be set, in order to analyze their trend.

Moreover, the visualization of each single contact force both in the grasp-state space graphs and in finger schemes can be activated by checking the relative button. In detail:

- the contact forces f_1, f_2, and f_3 are denoted respectively as yellow, orange, and blue surfaces in the grasp-state space graphs, while the green surfaces indicate the portion of the grasp-state spaces where the forces are all positive, therefore indicating a stable grasp. The green (stable grasp) or red (unstable grasp) point indicates the current configuration of the parameters (θ_2, k_2) for the two-phalanx finger or (θ_2, θ_3) and (k_2, k_3) for the three-phalanx finger;
- the vectors representing the contact forces in the finger schemes are green or red if the forces are, respectively, positive or negative. The blue vectors, instead, indicate the tangential forces acting at the object contact points.

Furthermore, the GUI language can be set (the current version only supports English and Italian).

- *Section 12—Results*

This section shows the main analytic outcomes obtained from the software: normal and tangential forces and values of the performance indexes both for two- and three-phalanx fingers. In the configuration considered in Figure 7, the grasped object, a disk, is positioned at the same distance from the base joint O_1 for the two finger architectures, but the grasp results stable only for the three-phalanx finger. This is due to the fact that, for the given combination of the chosen parameters, in the case of the two-phalanx finger the force f_2 is negative, while in the case of the three-phalanx finger all the forces are positive. Specifically:

- as can be seen in the grasp-state space graph of the two-phalanx finger (Figure 8a), the force f_2—orange surface—is always negative for each value of the (θ_2, k_2) parameters, so that the only way to obtain a stable grasp is that of changing the other parameters, such as the phalanx length or thickness, the friction coefficients, or the springs features;
- in the case of the three-phalanx finger (Figure 8b) the first grasp-state space, which is a function of the (θ_2, θ_3) parameters, shows a stable grasp—green surface—just in the 38.2% of the defined space, mainly due to the trend of f_2 surface; on the other hand, the second grasp-state space, which is a function of the (k_2, k_3) parameters, shows a stable grasp in the 91.2% of the defined space.

As an example, if in the case of the three-phalanx finger the friction coefficients are modified from $\mu_1 = \mu_2 = \mu_3 = 0.8$ to $\mu_1 = \mu_2 = \mu_3 = 0.6$ the grasp becomes unstable (as shown in Figure 9); this result highlights the importance of friction in the grasp stability problem and it also shows how this software could be useful in finding the best design parameters for an efficient underactuated gripper.

The proposed toolbox helped in the design choices of the Adam's Hand prototype family that is shown in Figure 10. It was especially useful in setting the stiffness of the joint springs to increase the stable portion of the grasp-state space of Adam's Hand.

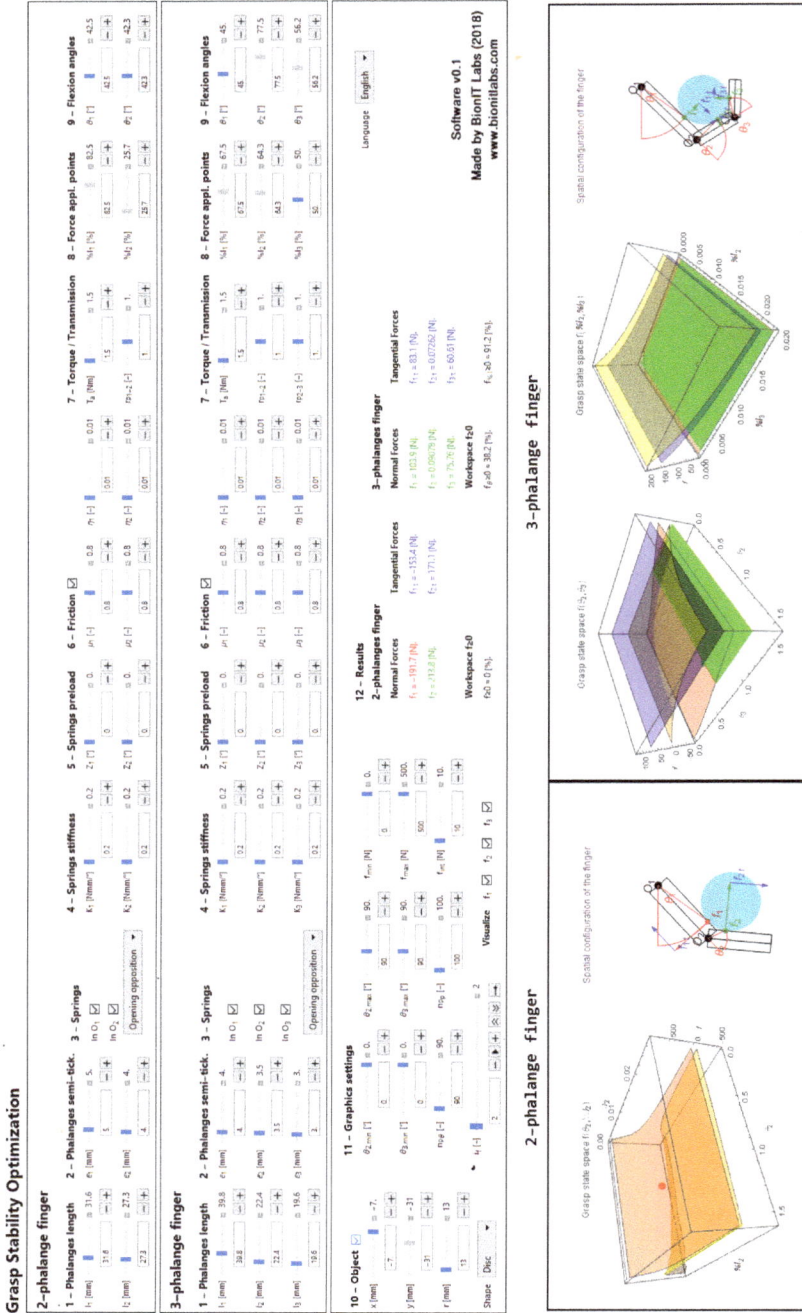

Figure 7. Graphical user interface (GUI) of the developed software. Please refer to the online colored version for a better view.

2-phalange finger

Grasp state space f(θ_2, $\%l_2$)

Spatial configuration of the finger

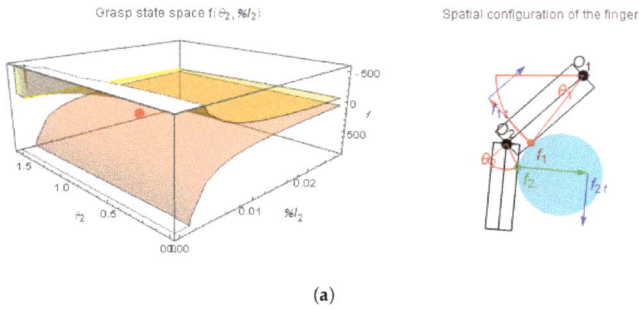

(**a**)

3-phalange finger

Grasp state space f(θ_2, θ_3)

Grasp state space f($\%l_2$, $\%l_3$)

Spatial configuration of the finger

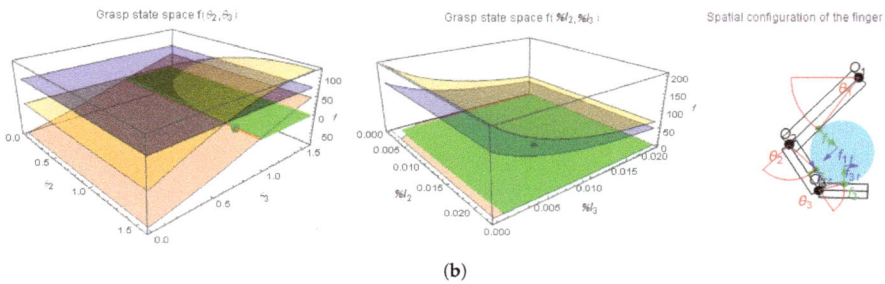

(**b**)

Figure 8. Grasp-state space graphs and finger schemes of the two-phalanx finger (**a**) and of the three-phalanx finger (**b**).

3-phalange finger

Grasp state space f(θ_2, θ_3)

Grasp state space f($\%l_2$, $\%l_3$)

Spatial configuration of the finger

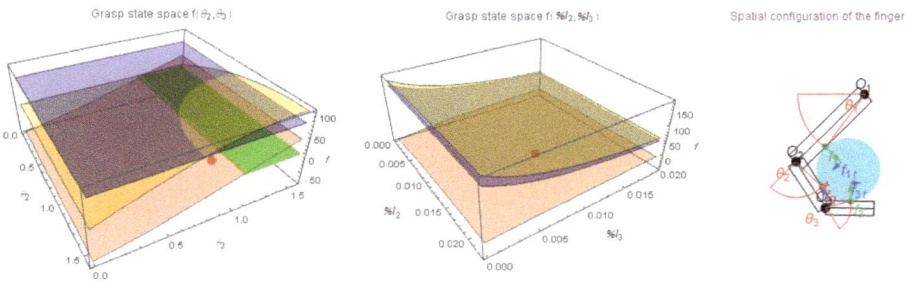

Figure 9. Grasp-state space graphs and configuration of the three-phalanx finger for an unstable grasp due to insufficient friction.

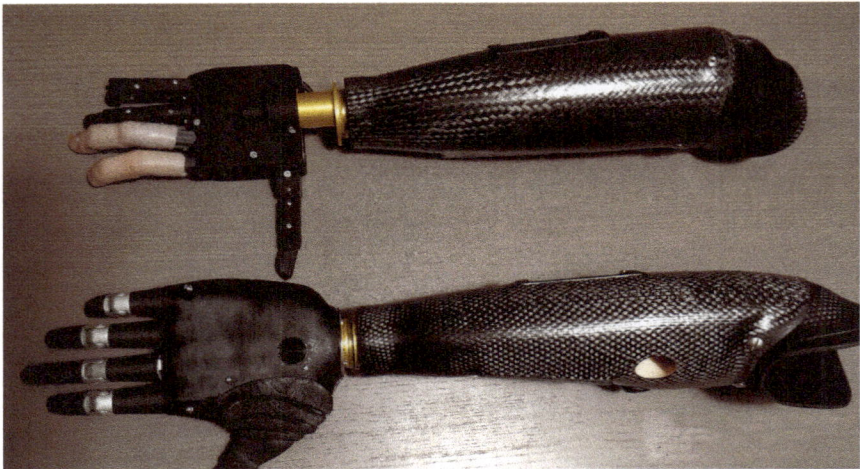

Figure 10. Adam's Hand prototype family: alpha-prototype at the top and beta-prototype at the bottom of the image.

5. Conclusions

In this paper, a software framework was presented to help a user during the design stage of a humanoid robotic hand that employs underactuated fingers towards grasp stability optimization. The tool is highly parameterized to cope with various parameters that include phalanx thickness and length, friction, joint spring properties, and driving torque. Although it was primarily intended for a parametric design of a humanoid underactuated hand using simulated grasping, it could also be valuable as an educational tool to help non-expert users or students to understand the principles underlying underactuated grasp by visualizing how the parameter slide bars impact on the stability indexes.

Future developments will be devoted to extend the single point contact model by also taking into account the linear and circular contact. The interaction with the grasped object by more than one finger at a time will be added to the system. Efforts will be made to include deformability of the phalanges, fingers, and grasped object towards a fully soft underactuated design. Finally, while many quality measures for grasps have been proposed in the literature, the use of these measures for automatic grasp choice remains an open issue [17]. Therefore, grasp quality metrics other than stability will be considered, for example, by taking into account the task requirement and following knowledge-based approaches.

Author Contributions: Conceptualization, G.A.Z., G.R. and A.M.; Investigation G.A.Z., G.R. and A.M.; Methodology, G.A.Z., G.R. and A.M.; G.A.Z., G.R. and A.M. wrote the paper.

Funding: The research leading to these results has received funding from the Horizon 2020 European Commission under grant agreement n. 821988 ADE

Conflicts of Interest: The authors declare no conflict of interest.

References

1. Russo, M.; Ceccarelli, M.; Corves, B.; Hüsing, M.; Lorenz, M.; Cafolla, D.; Carbone, G. Design and Test of a Gripper Prototype for Horticulture Products. *Robot. Comput. Integr. Manuf.* **2017**, *44*, 266–275. [CrossRef]
2. Yao, S.; Ceccarelli, M.; Carbone, G.; Dong, Z. Grasp configuration planning for a low-cost and easy-operation underactuated three-fingered robot hand. *Mech. Mach. Theory* **2018**, *129*, 51–69. [CrossRef]
3. Kragten, G.A.; Herder, J.L. The ability of underactuated hands to grasp and hold objects. *Mech. Mach. Theory* **2010**, *43*, 408–425. [CrossRef]

4.   Miller, A.; Allen, P. GraspIt! *IEEE Robot. Autom. Mag.* **2004**, *11*, 110–122. [CrossRef]
5.   Leon, B.; Ulbrich, S.; Diankov, R.; Puche, G.; Przybylski, M.; Morales, A.; Asfour, T.; Moisio, S.; Bohg, J.; Kuffner, J.; et al. Opengrasp: A toolkit for robot grasping simulation. *Simul. Model. Program. Auton. Robot.* **2010**, *11*, 109–120.
6.   Aukes, D.; Cutkosky, M. Simulation-based tools for evaluating underactuated hand designs. In Proceedings of the 2013 IEEE International Conference on Robotics and Automation (ICRA 2013), Karlsruhe, Germany, 6–10 May 2013; pp. 2067–2073. [CrossRef]
7.   Malvezzi, M.; Gioioso, G.; Salvietti, G.; Prattichizzo, D. SynGrasp: A MATLAB Toolbox for Underactuated and Compliant Hands. *Robot. Autom. Mag. IEEE* **2015**, *22*, 52–68. [CrossRef]
8.   Calli, B.; Walsman, A.; Singh, A.; Srinivasa, S.; Abbeel, P.; Dollar, A.M. Benchmarking in Manipulation Research: The YCB Object and Model Set and Benchmarking Protocols. *IEEE Robot. Autom. Mag.* **2015**, *22*, 36–52. [CrossRef]
9.   Adam's Hand. Adam's Hand Website. Available online: http://www.adamshand.it (accessed on 27 December 2018).
10. Zappatore, G.A.; Reina, G.; Messina, A. Adam's Hand: An Underactuated Robotic End-Effector. In *Advances in Italian Mechanism Science*; Springer: Cham, Switzerland, 2016; pp. 239–246.
11. Zappatore, G.A.; Reina, G.; Messina, A. Analysis of a highly underactuated robotic hand. *Int. J. Mech. Control* **2017**, *18*, 17–24.
12. Zappatore, G.A.; Reina, G.; Messina, A. A proposed software framework for studying the grasp stability of underactuated fingers. In *Mechanism Design for Robotics, Proceedings of the IFToMM Symposium on Mechanism Design for Robotics, Udine, Italy, 11–13 September 2018*; Springer: Cham, Switzerland, 2018; pp. 202–210.
13. Birglen, L.; Laliberté, T.; Gosselin, C. *Underactuated Robotic Hands*; Springer: Berlin/Heidelberg, Germany, 2008.
14. Wolfram, S. *Mathematica: A System for Doing Mathematics by Computer*; Wolfram Research Inc.: New York, NY, USA, 1988.
15. Buryanov, A.; Kotiuk, V. Proportions of Hand Segments. *Int. J. Morphol.* **2010**, *28*, 755–758. [CrossRef]
16. Oberg, E.; Jones, F.D.; Horton, H.L.; Ryffell, H.H.; Heald, R.M.; McCauley, C.J. *Machinery's Handbook*; Industrial Press: South Norwalk, CT, USA, 2000.
17. Okamura, A.; Smaby, N.; Cutkosky, M.R. An overview of dexterous manipulation. In Proceedings of the IEEE International Conference on Robotics and Automation, San Francisco, CA, USA, 24–28 April 2000; pp. 255–263.

robotics

MDPI

Article

On the Design of a Safe Human-Friendly Teleoperated System for Doppler Sonography

Juan Sebastián Sandoval Arévalo *, Med Amine Laribi, Saïd Zeghloul and Marc Arsicault

Department of GMSC, Pprime Institute, CNRS, ENSMA, University of Poitiers, UPR 3346 Poitiers, France;
med.amine.laribi@univ-poitiers.fr (M.A.L.); said.zeghloul@univ-poitiers.fr (S.Z.);
marc.arsicault@univ-poitiers.fr (M.A.)
* Correspondence: juan.sebastian.sandoval.arevalo@univ-poitiers.fr; Tel.: +33-5-4949-6538

Received: 11 January 2019; Accepted: 12 April 2019; Published: 15 April 2019

Abstract: Variable stiffness actuators are employed to improve the safety features of robots that share a common workspace with humans. In this paper, a study of a joint variable stiffness device developed by PPRIME Institute—called V2SOM— for implementation in the joints of a multi-DoF robot is presented. A comparison of the interaction forces produced by a rigid body robot and a flexible robot using the V2SOM is provided through a dynamic simulator of a 7-DoF robot. As an example of potential applications, robot-assisted Doppler echography is proposed, which mainly focuses on guaranteeing patient safety when the robot holding the ultrasound probe comes into contact with the patient. For this purpose, an evaluation of both joint and Cartesian control approaches is provided. The simulation results allow us to corroborate the effectiveness of the V2SOM device to guarantee human safety when it is implemented in a multi-DoF robot.

Keywords: safe physical human–robot interaction (pHRI); variable stiffness actuator (VSA); collaborative robots; robot-assisted Doppler sonography

1. Introduction

The capability of industrial robots to execute tasks significantly faster than humans has improved the efficiency of several industrial processes. However, there exist numerous tasks that are harder to automate, where human execution is required. The use of Cobots (i.e., collaborative robots) appears to be an effective solution to improve the execution of complex tasks where humans are required. Unlike the classical industrial robots, which are usually isolated and avoid physical contact with humans, Cobots share a common workspace with humans and cooperate with them to achieve a desired task [1].

At this time, the use of collaborative robots in medical and industrial applications is rapidly growing. In the context of robot-assisted Doppler sonography, a teleoperated manipulator holds an ultrasound probe and reproduces the same movements over the patient, which are executed by the medical expert manipulating a fictive probe from a master site. In this application, the efforts applied by the manipulator over the patient must be regulated to ensure patient safety and thus, it is important to create compliance in the robot movements.

When using Cobots, the most important issue is to guarantee a safe human–robot coexistence. In this regard, several solutions have been studied [2]. Park et al. led the use of a viscoelastic casing in the robot's body to reduce consequences of any impact [3]. Fritzsche et al. proposed the supervision of the impact forces by covering the robot's body with tactile sensors [4]. Human safety can also be ensured by providing the robot with compliant motion capabilities. Two main strategies are defined for this purpose. The first one concerns the use of specific control approaches in order to provide the robot with compliant motions, such as the well-known impedance control [5], admittance control [6] or the compliance control [7] approaches. Some of them react to the external forces applied to the robot, which are either measured by a force/torque sensor or estimated by a disturbance observer [8].

Furthermore, more simple controllers do not need an external torque measure or estimation, such as the one proposed in [7], which uses the potential energy of a virtual spring to perform the compliance motion. On the other hand, a second strategy implementing compliant motion capabilities involves the use of joint compliant mechanisms, allowing us to introduce intrinsic compliance to the robot. For instance, series elastic actuators (SEA) are simple compliant mechanisms, including a mechanical spring between the motor transmission output and the robot link [9]. Unlike the SEA that proposes a constant stiffness, variable stiffness actuators (VSA) are capable of providing adjustable stiffness values according to the requirements [10,11]. Several VSAs have been proposed since the 1980s. Some of them propose mechanical solutions for modifying the stiffness, such as the use of leaf springs [12,13]. Other more complex mechanisms include an extra actuator to vary the stiffness behavior [14,15]. This is also the case for the V2SOM mechanism, a novel rotational variable stiffness actuator that was recently presented by PPRIME Institute. This mechanism presents an innovative stiffness behavior, which is smoothened in the vicinity of zero deflection through the use of a cam-follower mechanism. In the case of collision, stiffness sharply sinks to a steady constant torque threshold, which is tunable according to the load variation [16]. The working principle of V2SOM and an evaluation of its safety performance when implemented on a multi-DoF is presented in this paper.

Various human safety indexes have been studied to validate the effectiveness of the compliant mechanism, such as the head injury criterion (HIC) [17] or the head impact power (HIP) criterion [18], which are used when evaluating the consequences of an impact to the head. These criteria measure the displacements, velocities and accelerations of the head during crash impact tests [19] and are usually used in the automotive sector. Nevertheless, other safety criteria can be employed in robotics, such as the measures of displacements, velocities or accelerations. Furthermore, the measure/estimation of the interaction forces provides significant information for studying the safety behavior of a compliant mechanism.

The validation of safety performance for VSA is typically performed by studying a single-DoF system case. Thus, the collisions between the link attached to the VSA and the environment representing a human being are typically produced and evaluated. Nevertheless, this type of study only provides information about the local safety performance of the compliant mechanism and makes it impossible to evaluate its global performance when using it in a multi-DoF robot.

In this paper, a study of the V2SOM safety performance is presented. Unlike classical studies evaluating the safety performance in a single-DoF model, the presented work studies the performance of a 7-DoF robot using V2SOM on each joint. For this purpose, the dynamic model of a commercial 7-DoF robot has been modified to include the compliant mechanism on each joint. Moreover, two study cases are presented. The first one considers the execution of desired trajectories in joint spaces. In the second one, an application, namely robot-assisted Doppler sonography, is considered. Finally, interaction forces are evaluated according to the human safety index since low accelerations are performed in the application.

This paper is organized as follows. In Section 2, the modeling of a multi-DoF robot using the V2SOM is depicted. Furthermore, joint and cartesian control approaches for executing tracking trajectory tasks are explained. In Section 3, the robot-assisted Doppler sonography application and their issues are presented. A study case for the comparison of the safety performance of a rigid body robot and a robot using the V2SOM on each joint is also presented. A discussion of the obtained results and the conclusions of the presented work are provided at the end of this present paper.

2. Materials and Methods

In the following section, the dynamic model of a multi-DoF robot with joint flexibility that is provided by the implementation of the V2SOM is described. First, the working principle of the V2SOM is depicted. Subsequently, the dynamic robot model with the V2SOM for the execution of tracking trajectory tasks in both the joint and cartesian space is presented.

2.1. Working Principle of V2SOM

In order to ensure safe behavior in the event of a collision, a mechanism has been designed to provide a finite torque's slope when approaching to the zero deflection. When the deflection increases, the stiffness smoothly decreases until it reaches a threshold torque level T_{max}. The performance curve of the V2SOM is given by:

$$T_\theta = T_{max}\left(1 - e^{-s\theta}\right), \tag{1}$$

where s is a positive constant value and θ represents the elastic deflection angle.

In general, two working modes are identified for the V2SOM. The transition between the two modes smoothly occurs in the case of a strong collision, as illustrated in Figure 1a. The first mode represents a high stiffness behavior (I) and is defined within a deflection range $[0, \theta_1]$ and a torque range $[0, T_1]$, respectively. The value of T_1 defines the limit of normal torque working conditions. When this value is exceeded, the impact absorbing mode (II) is activated, which is characterized by a progressive decrease in the stiffness before reaching the torque threshold T_{max}. In the developed V2SOM prototype, the deflection θ_1 corresponds to a torque $T_1 = 0.8 \cdot T_{max}$ and supports a maximum deflection $\theta_{max} = \frac{\pi}{2}$.

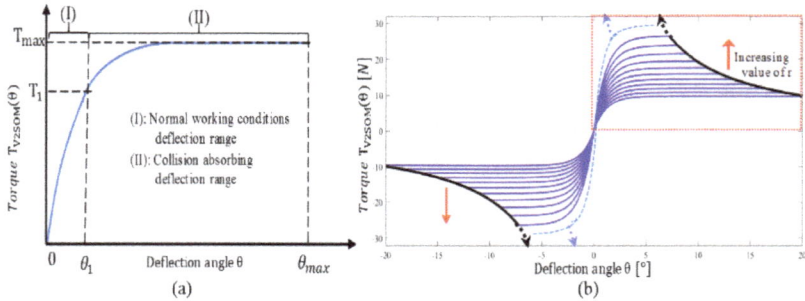

Figure 1. (**a**) Working modes of the V2SOM; and (**b**) Various performance curves of the V2SOM according to different values of r (tunable reduction ratio).

The nonlinear behavior is achieved through a cam/follower system. Two blocks form the mechanics of the V2SOM, with each one having a specific task. The two blocks are rigidly coupled as it can be corroborated in the kinematic scheme of the Figure 2a.

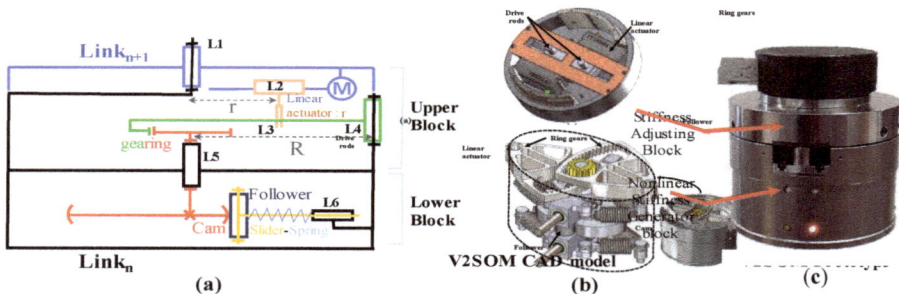

Figure 2. (**a**) Kinematic scheme; (**b**) CAD model of the upper block (top) and lower block (bottom); and (**c**) First prototype of V2SOM.

The upper block, called the stiffness adjusting block, is basically a deflection angle reducer (torque amplifier) with a tunable reduction ratio. This ratio can continuously be adjusted by modifying the reducer's tuning parameter r through the actuated joint $L2$. Each value of r leads to a different torque curve with its corresponding T_{max} and θ_{max} values. Some of these curves are shown in Figure 1b. Moreover, the V2SOM is designed to have a symmetric torque vs. deflection behavior, which allows it to work in two rotation directions. The lower block, called the nonlinear stiffness generator block, is based on a cam/follower mechanism with some additional springs. The cam profile, related to a given positive parameter s, generates the desired torque curve T_θ vs. deflection θ (see Equation (1)). The CAD design and an image of the first prototype are shown in Figure 2b,c, respectively. Readers are invited to refer to [20] in order to obtain further details about the design phase of the V2SOM.

In the following section, a description of the dynamic model of a multi-DoF robot with joint flexibility by means of the V2SOM implementation on its joints is presented. The two control designs for executing tasks in the joint and cartesian levels are also depicted.

2.2. Joint Control Design

The dynamic modeling of a flexible joint proposed by [21] is useful for representing a n-DoF serial robot implementing V2SOM on each joint as follows:

$$M(q_o)\ddot{q}_o + C\left(q_o, \dot{q}_o\right)\dot{q}_o + g(q_o) = T_\theta + T_{ext}, \tag{2}$$

The vector $q_o \in \mathfrak{R}^n$ comprises the link side positions and $T_\theta \in \mathfrak{R}^n$ contains the output torque provided by V2SOM whose behavior is explained by Equation (1), as shown in Figure 2. The external forces acting on the robot are represented by the torque vector $T_{ext} \in \mathfrak{R}^n$. Moreover, the motor side dynamics is defined by:

$$B\ddot{q}_i + T_\theta = T_i - T_f, \tag{3}$$

where $B \in \mathfrak{R}^{n\times n}$ is the motor inertia matrix, $q_i \in \mathfrak{R}^n$ is the vector containing the motor side positions and $T_i \in \mathfrak{R}^n$ contains the motor torques. The friction torques are represented by $T_f \in \mathfrak{R}^n$. After this, the elastic deflection angle is defined by $\theta = q_i - q_o$.

Several control approaches can be used to control a robot, including flexible joints, such as the one presented in [22]. This approach was proposed for fast movements, such as pick-and-place applications. In this case, in order to execute joint tracking trajectory tasks, the torque motor T_i can be controlled through a PD regulator that is added to a gravity compensation term in a similar way to the approach proposed in [23]:

$$T_i = K_p(q_d - q_i) - K_d\dot{q}_i + \hat{g}(q_o), \tag{4}$$

The vector $q_d \in \mathfrak{R}^n$ is the desired link side position. Figure 3 shows the block diagram of the implemented control approach. Furthermore, the passivity of the system can be guaranteed by properly choosing the constant values K_p and K_d.

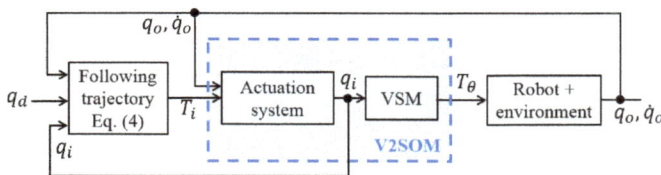

Figure 3. Block diagram of the proposed joint space control architecture.

2.3. Cartesian Control Design

In the case of cartesian tracking trajectory tasks of dimension m, the torque motor T_i can also be achieved through a PD regulator and a gravity compensation term, as follows:

$$T_i = J^T\left[K_{p_x}(x_d - x_i) - K_{d_x}\dot{x}_i\right] - N(q)\xi + \hat{g}(q_o), \tag{5}$$

where $x_d \in \mathfrak{R}^m$ is the vector of the desired end effector positions. Similar to the joint space case, the passivity of the system can be ensured with a proper choice of the constant values K_p and K_d. $J(q) \in \mathfrak{R}^{m \times n}$ is the Jacobian matrix and $N(q) = I - J^T J^{+T}$ is a null space projector that allows us to optimize an objective function represented by ξ. Figure 4 shows the block diagram of the implemented control approach for the cartesian case.

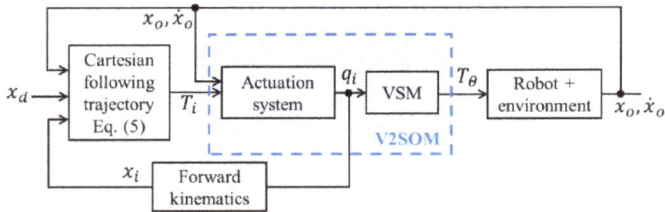

Figure 4. Block diagram representing the proposed cartesian space control architecture.

3. Implementation and Results

In this section, the robot-assisted Doppler sonography is first presented. As mentioned above, the use of V2SOM can be useful for this medical application. Subsequently, two study cases allowing us to compare the safety performance between a rigid body and a compliant robot using the V2SOM are presented, where the latter study case concerns the mentioned medical application.

3.1. Robot-Assisted Doppler Sonography

Several studies demonstrate the appearance of work-related musculoskeletal disorders (WRMD) due to the uncomfortable postures adopted by the sonographers during the examinations [24,25]. The motion capture analysis performed during examinations has allowed us to confirm that sonographers frequently take postures completely out of the comfort zone. In order to avoid the existence of WRMD, a robotized platform for Doppler sonography has been proposed by PPRIME Institute. The medical expert, located at the master site, operates a 3DoF haptic device (Figure 5). In a real-life scenario, the sonographer will perform examinations on patients lying on the examination table in the same conditions as in his medical office. The setup shown in Figure 5 is used to perform the first experimental tests. The haptic device pedals the movements of the serial robot and maintains the ultrasound probe over the patient. The use of the master device instead of manually manipulating the probe allows medical experts to restrict their movements to the comfort zone.

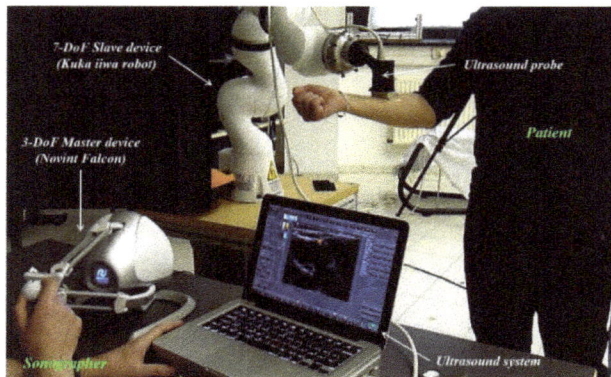

Figure 5. Teleoperated system for Doppler echography.

In order to ensure safe patient–robot contact, it is important to create compliant behavior in the slave robot. Therefore, compliance can be provided by implementing V2SOM on each joint of the robot. A study case is provided below for a classical trajectory executed at the beginning of a robot-assisted Doppler sonography examination.

3.2. Study Cases

Hereafter, some preliminary results of the safety performance of a multi-DoF collaborative robot using V2SOM to provide it with joint flexibility are presented. A 7-DoF Kuka IIWA robot has been used for this experience. This is one of the most relevant torque-controlled collaborative robots that are employed by the research community.

3.2.1. Joint Space Trajectory Task

For this first experience, the cartesian workspace is restrained to the YZ plane. According to this restriction, only the movements on joints 2 and 4 are activated, while the rest of the joints have been blocked to fixed joint position values. The link side position vector is defined as $q_o = \left\{ \pi/2,\ q_{o_2} - \pi/2,\ 0,\ q_{o_4},\ 0,\ 0,\ 0 \right\}$.

After this, the proposed V2SOM is implemented in the second and fourth joints with the purpose to provide safe compliant behavior in the case of a collision with an external object. A joint desired linear trajectory for the joint 2 is defined, i.e., q_{d_2} from $0°$ to $90°$. For joint 4, a fixed desired angle $q_{d_4} = 0°$ was set. A compliant object is placed on the robot's workspace, which interferes with its trajectory. The external compliant object is characterized by certain stiffness and damping values of $k_e = 1000$ N/m and $d_e = 10$ Ns^2/m, respectively. Figure 6a illustrates the proposed study case.

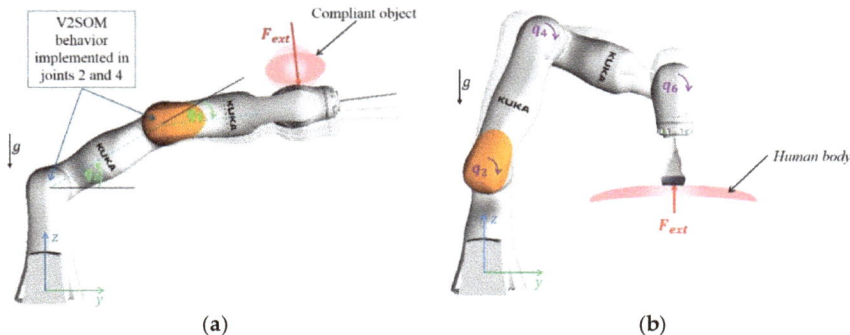

Figure 6. (**a**) First study case: a 7-DoF robot using V2SOM in joints 2 and 4 executes a trajectory within a planar workspace while a compliant object obstructs with the desired trajectory. (**b**) Second Study case: a 7-DoF robot integrating V2SOM on each joint executes a 3D trajectory until getting in contact with the human body.

The constant parameters were defined as follows: $K_p = diag\{4000,\ 12000,\ 4000,\ 12000,\ 4000,\ 4000,\ 4000\}$ and $K_d = diag\{300,\ 811,\ 300,\ 811,\ 300,\ 300,\ 0.15\}$. Friction effects were neglected for the sake of simplicity, i.e., $T_f \approx 0$. Furthermore, the compliance parameters of the V2SOM for joints 2 and 4 were selected as follows: $T_{max} = \pm 100$ Nm and $s = -184.428$ rad^{-1} for a deflection value $\theta_1 = 0.5°$. Figures 7 and 8 show the obtained simulation results for this study case. A comparison between the robot's behavior in two different compliance configurations is presented: when using a rigid body robot (i.e., $T_\theta \approx T_i$) and when V2SOM is implemented in joints 2 and 4. Considering the low velocities used in these examinations, the HIC and HIP criteria are not suitable for evaluating the safety performance of the robot. Thus, the interaction forces F_{ext} induced for the two configurations are proposed for the safety performance index.

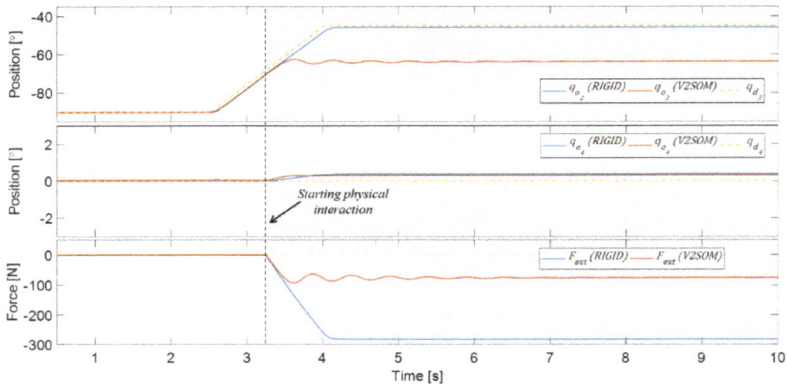

Figure 7. (**Top**) Position signals for joints 2 and 4 in the two cases: with ("V2SOM") and without ("RIGID") using the V2SOM. (**Bottom**) Interaction force F_{ext} generated in these two cases.

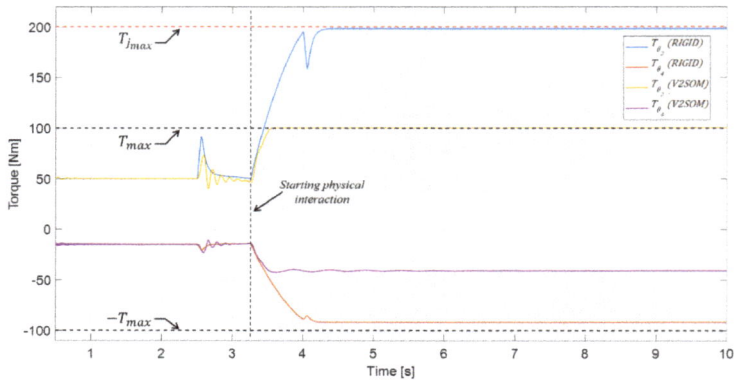

Figure 8. Link side torque signals $T_{\theta_{2,4}}$ generated by joints 2 and 4 in both cases: when using a rigid body robot and when the output torques are provided by the implemented V2SOMs.

The interaction force F_{ext} generated during the physical contact between the robot and the compliant object naturally induces a variation in the joint position signals. In the case of a rigid body robot implementing a classical joint position control strategy, the PD regulator, directed by the constant values K_p and K_d, are forced to reach the desired trajectory using the maximum torque provided by the motor. This behavior can be verified in Figures 7 and 8, where joint 2 reaches its motor torque limit $T_{j_{max}}$ (± 200 Nm according to the manufacturer) that provides a joint trajectory close to the desired one. As expected, this behavior also increases the magnitude of the interaction force (Figure 7, bottom). On the other hand, the intrinsic compliance provided by the proposed mechanism induces a decrease in the interaction force, which naturally causes a loss of accuracy in the following trajectory task. Moreover, it can be verified in Figure 8 that the variable stiffness law provided by V2SOM limits the motor torques to the defined torque threshold $T_{max} = \pm 100$ Nm. It is worth mentioning that the torque variations generated from $t = 2.5$ s correspond to the variation of the desired trajectory for joint 2.

3.2.2. Cartesian Space Trajectory Task for Doppler Sonography

In the context of robot-assisted Doppler sonography, the robot holding the probe is first positioned at its zero position (the probe is placed over the patient). After this, the robot executes a vertical trajectory until it comes into contact with the patient's body. This vertical trajectory corresponds to the first part of the proposed desired 3D cartesian trajectory, which is denoted as $x_d(t)$. During the second part, the robot moves away from the patient's body. In real life conditions, once this trajectory is executed, the sonographer is able to teleoperate the robot using the haptic device.

For this study case, the patient's body is represented as a compliant entity characterized by certain stiffness $k_e = 3000$ N/m and damping $d_e = 30$ Ns^2/m values, respectively. Figure 6b illustrates the proposed study case.

Constant parameters have been fixed as: $K_{p_x} = diag\{6000, 6000, 6000\}$ and $K_{d_x} = diag\{300, 300, 300\}$. Similar to the previous study case, friction effects have been neglected (i.e., $T_f \approx 0$) and the compliance parameters of the V2SOM were selected as: $T_{max} = \pm 30$ Nm and $s = -184.428$ rad^{-1} for a deflection value $\theta_1 = 0.5°$. The objective function was employed to stabilize the internal motion by reducing the joint velocities, i.e., $\xi = -0.1\dot{q}_i$.

The robot's behavior in three different configurations are compared: when using a rigid body robot, i.e., $T_\theta \approx T_i$; when a linear compliant behavior is implemented (constant stiffness); and when the V2SOM is implemented on each joint of the robot. The obtained results are shown in Figures 9 and 10.

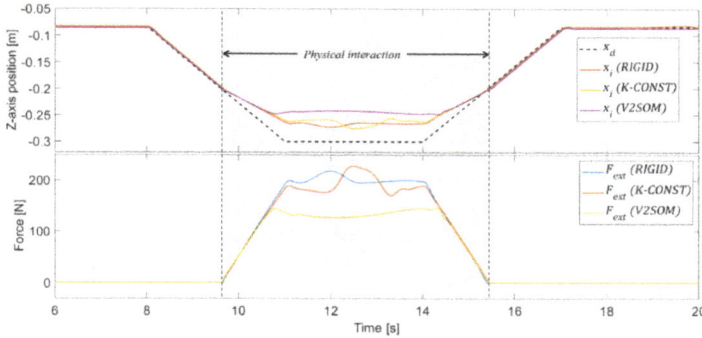

Figure 9. (**Top**) Current position along the z-axis; and (**Bottom**) Interaction force F_{ext} measured for the three configurations: rigid body, constant stiffness and when using the V2SOM.

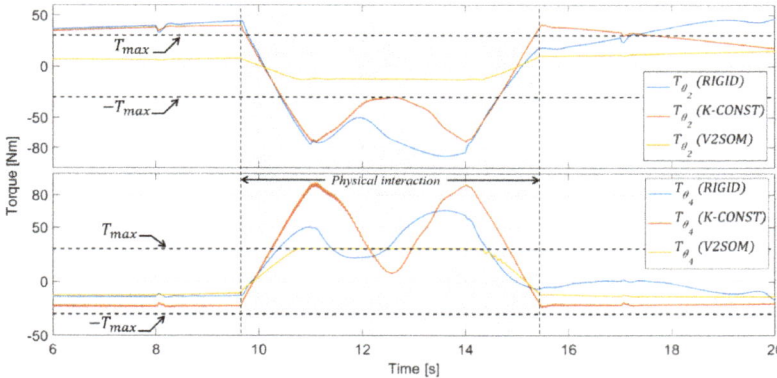

Figure 10. Link side torque signals T_θ in each compliant joint for the three different configurations: rigid body, constant stiffness and when using the V2SOM.

As expected for the medical application, the compliance behavior provided by the V2SOM decreases the interaction forces (Figure 9 bottom). Furthermore, the z-axis trajectory performed by the robot using V2SOM does not affect the quality of the Doppler test since the contact between the probe and human body is established and guaranteed. Figure 10 proves that the output torques provided by

V2SOM are always restricted by the desired limits $T_{max} = \pm 30$ Nm unlike the motor torque signals of the rigid body and constant stiffness cases.

4. Discussion

In the first study case, Figure 7 shows the position signals of joint 2 and the interaction force induced by the collision between the robot's wrist and the compliant object in the two configurations. The magnitude of this interaction force represents a safety index, which indicates a more human-friendly behavior when low values are measured. In this case, Figure 7 allows us to verify that the interaction force F_{ext} has significantly decreased for the configuration when using V2SOM in joints 2 and 4, which proves its safer performance.

Figure 8 shows the control torque signals for joints 2 and 4 in the two configurations. It is possible to verify that the output torque T_θ provided by the V2SOM that is implemented in joint 2 is constrained by the torque threshold T_{max} at a time of around 3.6 s. In contrast, the torque signals of the rigid body configuration are only restricted by the motor torque limits $T_{j_{max}}$.

In the second study case, the current position signals on the z-axis and the interaction force generated by the collision between the probe and the patient's body for the three configurations are shown in Figure 9. Similar to the previous case, it is shown that the interaction force F_{ext} has considerably decreased when using the V2SOM, which demonstrates safer performance compared to the rigid body robot configuration.

Although the external efforts generated during the physical interaction are experienced by all the robot joints, Figure 10 allows us to verify that the efforts felt by joint 4 are particularly important as they reach the torque threshold T_{max} of 30 Nm. An opposing case can be seen for the rigid body and the linear compliance configurations as the only restriction imposed to the joint torques concerns the motor torque limits, which are not reached for this case.

It is worth mentioning that the use of V2SOM intrinsically improves the safety performance of the multi-DoF robot used in the teleoperation system, as verified by the interaction forces in the presented study cases. Further details about the safety performance evaluation of V2SOM through an evaluation of safety criteria can be found in [26,27].

5. Conclusions

In this paper, the safety performance of a collaborative robot using the V2SOM, a variable stiffness mechanism conceived by PPRIME Institute, has been presented. A dynamic simulator to integrate the nonlinear compliant behavior of the V2SOM to the joints of a 7-DoF collaborative robot (i.e., a Kuka IIWA robot) has been developed. The dynamic robot's model has been modified to include the V2SOM compliance model. Two study cases were proposed to evaluate the safety performance of the modeling system. Firstly, the physical interaction between the robot and a compliant object when executing joint space trajectories was studied. Secondly, the use of the modified robot model for a medical application, namely robot-assisted Doppler sonography, was presented, where the forces applied by the robot holding the ultrasound probe over the patient must be minimized. Several comparisons were made in terms of the safety performance between the robot using and without using the V2SOM which mainly considers the generated interaction forces as a consistent safety criterion. This provides evidence of a safer response when the V2SOM is implemented.

Author Contributions: J.S.S.A. designed the experiments and co-wrote the paper with M.A.L. The research work was supervised by S.Z. and M.A.

Funding: This research was funded by the French National Research Agency, convention ANR-14-CE27-0016, under the ANR project SISCob "Safety Intelligent Sensor for Cobots". This research was also supported by the French region "Nouvelle-Aquitaine" (program HABISAN 2015-2020) with the financial participation of the European Union (FEDER/ERDF, European Regional Development Fund).

Acknowledgments: This work was sponsored by the French government research program *Investissements d'avenir* through the Robotex Equipment of Excellence (ANR-10-EQPX-44).

Robotics **2019**, *8*, 29

Conflicts of Interest: The authors declare no conflict of interest.

References

1. Gillespie, R.B.; Colgate, J.E.; Peshkin, M.A. A general framework for cobot control. *IEEE Trans. Rob. Autom.* **2001**, *17*, 391–401. [CrossRef]
2. De Santis, A.; Siciliano, B.; De Luca, A.; Bicchi, A. An atlas of physical human–robot interaction. *Mech. Mach. Theory* **2008**, *43*, 253–270. [CrossRef]
3. Park, J.J.; Haddadin, S.; Song, J.B.; Albu-Schäffer, A. Designing optimally safe robot surface properties for minimizing the stress characteristics of human-robot collisions. In Proceedings of the 2011 IEEE International Conference on Robotics and Automation, Shanghai, China, 9–13 May 2011; pp. 5413–5420.
4. Fritzsche, M.; Elkmann, N.; Schulenburg, E. Tactile sensing: A key technology for safe physical human robot interaction. In Proceedings of the 6th International Conference on Human-Robot Interaction (HRI '11), Lausanne, Switzerland, 6–9 March 2011; pp. 139–140.
5. Chiaverini, S.; Siciliano, B.; Villani, L. A survey of robot interaction control schemes with experimental comparison. *IEEE/ASME Trans. Mechatron.* **1999**, *4*, 273–285. [CrossRef]
6. Ott, C.; Mukherjee, R.; Nakamura, Y. A hybrid system framework for unified impedance and admittance control. *J. Intell. Rob. Syst.* **2014**, *78*, 359–375. [CrossRef]
7. Dietrich, A.; Wimbock, T.; Albu-Schaffer, A.; Hirzinger, G. Integration of reactive, torque-based self-collision avoidance into a task hierarchy. *IEEE Trans. Rob.* **2012**, *28*, 1278–1293. [CrossRef]
8. Sadeghian, H.; Keshmiri, M.; Villani, L.; Siciliano, B. Null-space impedance control with disturbance observer. In Proceedings of the IEEE/RSJ International Conference on Intelligent Robots and Systems, Vila Moura, Portugal, 7–12 October 2012; pp. 2795–2800.
9. Pratt, G.A.; Williamson, M.M. Series elastic actuators. In Proceedings of the 1995 IEEE/RSJ International Conference on Human Robot Interaction and Cooperative Robots, Pittsburgh, PA, USA, 5–9 August 1995; pp. 399–406.
10. Bicchi, A.; Tonietti, G.; Bavaro, M.; Piccigallo, M. Variable stiffness actuators for fast and safe motion control. *Rob. Res.* **2005**, *15*, 527–536.
11. Grioli, G.; Wolf, S.; Garabini, M.; Catalano, M.; Burdet, E.; Caldwell, D.; Carloni, R.; Friedl, W.; Grebenstein, M.; Laffranchi, M.; et al. Variable stiffness actuators: The user's point of view. *Int. J. Rob. Res.* **2015**, *34*, 727–743. [CrossRef]
12. Morita, T.; Sugano, S. Design and development of a new robot joint using a mechanical impedance adjuster. In Proceedings of the 1995 IEEE International Conference on Robotics and Automation, Nagoya, Japan, 21–27 May 1995; pp. 2469–2475.
13. Groothuis, S.; Carloni, R.; Stramigioli, S. A Novel Variable Stiffness Mechanism Capable of an Infinite Stiffness Range and Unlimited Decoupled Output Motion. *Actuators* **2014**, *3*, 107–123. [CrossRef]
14. Wolf, S.; Hirzinger, G. A new variable stiffness design: Matching requirements of the next robot generation. In Proceedings of the IEEE International Conference on Robotics and Automation (ICRA 2008), Pasadena, CA, USA, 19–23 May 2008; pp. 1741–1746.
15. Schiavi, R.; Grioli, G.; Sen, S.; Bicchi, A. VSA-II: A novel prototype of variable stiffness actuator for safe and performing robots interacting with humans. In Proceedings of the International Conference on Robotics and Automation (ICRA 2008), Pasadena, CA, USA, 19–23 May 2008; pp. 2171–2176.
16. Ayoubi, Y.; Laribi, M.A.; Arsicault, M.; Zeghloul, S.; Courreges, F. Mechanical device with variable compliance for rotary motion transmission. FR/IFBT17CNRCOB. 2017.
17. National Highway Traffic Safety Administration, Department of Transportation (DOT). *Occupant Crash Protection—Head Injury Criterion S6.2 of MVSS 571.208*; Docket 69-7, Notice 17; NHTSA: Washington, DC, USA, 1972.
18. Newman, J.A.; Shewchenko, N.; Welbourne, E. A proposed new biomechanical head injury assessment function—the maximum power index'. In Proceedings of the 44th Stapp Car Crash Conference, Atlanta, GA, USA, 6–8 November 2000. SAE Paper No. 2000-01-SC16.
19. López-Martínez, J.J.; García-Vallejo, D.D.; Giménez-Fernández, A.A.; Torres-Moreno, J.L. A Flexible Multibody Model of a Safety Robot Arm for Experimental Validation and Analysis of Design Parameters. *ASME. J. Comput. Nonlinear Dynam.* **2013**, *9*, 011003–011003-10.

20. Sandoval, J.; Laribi, M.A.; Zeghloul, S.; Arsicault, M.; Poisson, G. Safety performance of a variable stiffness actuator for collaborative robots. In *Advances in Service and Industrial Robotics*; RAAD 2018, Ed.; Mechanisms and Machine Science. (In press)

21. Spong, M. Modeling and control of elastic joint robots. *ASME J. Dyn. Syst. Meas. Contr.* **1987**, *109*, 310–319. [CrossRef]

22. Tonietti, G.; Schiavi, R.; Bicchi, A. Design and Control of a Variable Stiffness Actuator for Safe and Fast Physical Human/Robot Interaction. In Proceedings of the 2005 IEEE International Conference on Robotics and Automation, Barcelona, Spain, 18–22 April 2005; pp. 526–531.

23. Albu-Schäffer, A.; Ott, C.; Hirzinger, G. A Unified Passivity-based Control Framework for Position, Torque and Impedance Control of Flexible Joint Robots. *Int. J. Rob. Res.* **2007**, *26*, 23–39. [CrossRef]

24. Gill, H.; Allison, H. Work-related musculoskeletal disorders in ultrasound: Can you reduce risk? *Ultrasound* **2015**, *23*, 224–230. [CrossRef] [PubMed]

25. Village, J.; Trask, C. Ergonomic analysis of postural and muscular loads to diagnostic sonographers. *Int. J. Ind. Ergon.* **2007**, *37*, 781–789. [CrossRef]

26. Ayoubi, Y.; Laribi, M.A.; Zeghloul, S.; Arsicault, M. V2SOM: A Novel Safety Mechanism Dedicated to a Cobot's Rotary Joints. *Robotics* **2019**, *8*, 18. [CrossRef]

27. Ayoubi, Y.; Laribi, M.A.; Zeghloul, S.; Arsicault, M. Design of V2SOM: The Safety Mechanism for Cobot's Rotary Joints. In *Mechanism Design for Robotics. MEDER 2018. Mechanisms and Machine Science*; Gasparetto, A., Ceccarelli, M., Eds.; Springer: Cham, The Netherlands, 2019; Volume 66.

MDPI

St. Alban-Anlage 66

4052 Basel

Switzerland

Tel. +41 61 683 77 34

Fax +41 61 302 89 18

www.mdpi.com

Robotics Editorial Office

E-mail: robotics@mdpi.com

www.mdpi.com/journal/robotics

www.ingramcontent.com/pod-product-compliance
Lightning Source LLC
Chambersburg PA
CBHW051848210326
41597CB00033B/5814